NUREG-1409

I0482647

Backfitting Guidelines

U.S. Nuclear Regulatory Commission

Office for Analysis and Evaluation of Operational Data

D. P. Allison, J. M. Conran, and C. A. Trottier

NUREG-1409

Backfitting Guidelines

Manuscript Completed: June 1990
Date Published: July 1990

D. P. Allison, J. M. Conran, and C. A. Trottier

Office for Analysis and Evaluation of Operational Data
U.S. Nuclear Regulatory Commission
Washington, DC 20555

ABSTRACT

The backfitting process is the process by which the U.S. Nuclear Regulatory Commission (NRC) decides whether to issue new or revised requirements or staff positions to licensees of nuclear power reactor facilities. Backfitting is expected to occur and is an inherent part of the regulatory process. However, it is to be done only after formal, systematic review to ensure that changes are properly justified and suitably defined. Requirements for proper justification of backfits and information requests are provided by two NRC rules, Title 10 of the *Code of Federal Regulations*, Sections 50.109 and 50.54(f). Three types of backfits are recognized. Cost-justified substantial safety improvements require backfit analyses and findings of substantial safety improvement and justified costs. Two types of exceptions, compliance exceptions and adequate protection exceptions, do not require findings of substantial safety improvements and costs are not considered. However, they are still backfits and they require documented evaluations to support use of the exceptions. Information requests (as opposed to backfits) require an analysis of the burden to be imposed to ensure that they are justified in view of the potential safety significance of the information requested. NRC procedures on backfitting include the Charter of the Committee to Review Generic Requirements for generic communications and NRC Manual Chapter 0514 and individual office procedures for plant-specific communications. Considerable guidance has been developed, control mechanisms are in place, and training has been provided to NRC and industry personnel. The Director of the Office for Analysis and Evaluation of Operational Data is responsible for oversight of backfitting programs, including obtaining industry comments. Initiatives are under way to better explain the process and conduct further training for industry and NRC personnel. Further initiatives are being considered in response to industry comments obtained in a recent survey concerning the effects of the regulatory process on licensees.

Contents

		Page
Abstract		iii
Executive Summary		vii
1	Introduction	1
2	Discussion	2
	2.1 Nature and Types of Backfits	2
	2.1.1 Background	2
	2.1.2 Backfit Determination	3
	2.1.3 Justification for Imposing Backfits	3
	2.2 Information Requests	5
	2.3 Staff Process for Identifying and Imposing Generic Backfits	6
	2.4 Staff Process for Identifying and Imposing Plant-Specific Backfits	7
	2.5 Filing a Backfit Claim or Appeal	9
	2.6 Current Status	11
3	Questions and Answers on Backfitting	12
	3.1 Backfit Determination and Imposition	12
	3.2 Generic Backfits	14
	3.3 Plant-Specific Backfits	15
	3.4 Backfit Analysis	18
	3.5 Appeals	19
	3,6 General questions	19

Appendices

A   The 1985 Backfit Rule

B   The 1988 Backfit Rule

C   Committee To Review Generic Requirements Charter

D   NRC Manual Chapter 0514, NRC Program for Management of Plant-Specific Backfitting of Nuclear Power Plants

E   Background Information for CRGR Review of Gi–70 and Gi–94 Resolutions

F   Sample Backfit Discussion (taken from NRC Bulletin 90–01)

EXECUTIVE SUMMARY

The backfitting process is the process by which the U.S. Nuclear Regulatory Commission (NRC) decides whether to issue new or revised requirements or staff positions to licensees of nuclear power reactor facilities. Backfitting is expected to occur and is an inherent part of the regulatory process. However, it is to be done only after formal, systematic review to ensure that changes are properly justified and suitably defined. The requirements of this process are intended to ensure order, discipline, and predictability and to enhance optimal use of NRC staff and licensee resources.

Title 10 of the *Code of Federal Regulations*, Section 50.109 (10 CFR 50.109), contains the backfit rule, which the NRC revised in 1985 to provide specific guidance and standards for backfitting decisions. The 1985 rule and the NRC manual chapter that implemented it were vacated by the U.S. Court of Appeals in 1987. The court stated that the rule was ambiguous about whether economic costs would be considered in ensuring or redefining adequate protection for the public health and safety or the common defense and security. In 1988 the NRC issued an amended backfit rule that was again subjected to court review and was upheld. The amended rule states clearly that economic costs will not be considered in cases of ensuring, defining, or redefining adequate protection or in cases of ensuring compliance with NRC requirements or written licensee commitments.

The backfit rule applies to both generic backfits and plant-specific backfits for power reactors. It defines a backfit as a modification of or addition to plant systems, structures, components, procedures, organization, design approval, or manufacturing license that may result from the imposition of a new or amended rule or regulatory staff position that became effective after specific dates. The rule recognizes three types of backfits. For backfits that do not meet one of the exceptions discussed below, a backfit analysis is required and it must be determined, based on that analysis, that the backfit will provide a substantial increase in overall protection of the public health and safety (or common defense and security) and that the direct and indirect costs for the facility are justified in view of the increased protection. Two types of exceptions are recognized, involving compliance and adequate protection. Such exceptions are still backfits, but they are justified differently. A documented evaluation is required, which states the objectives and purpose of the backfit and the basis for invoking the exception.

The rule (10 CFR 50.54(f)) requiring licensee responses to both generic and plant-specific information requests was revised along with the backfit rule. The rule stipulates that, except for information sought to verify licensee compliance with the current licensing basis, the NRC must prepare the reasons for the request to ensure the burden imposed on licensees is justified in view of the potential safety significance of the issue to be addressed.

One of the controls on generic backfitting and generic information requests is review by the Committee to Review Generic Requirements (CRGR). This committee of senior managers from various NRC offices was established in November 1981. Its objectives include eliminating unnecessary burdens on licensees, reducing radiation exposure to workers while implementing requirements, and optimizing use of NRC and licensee resources to assure safe operation. Following its review of a proposed generic communication the CRGR recommends approval, revision, or disapproval to the NRC Executive Director for Operations (EDO). If the office proposing the communication does not agree with the CRGR recommendation it may refer the issue to the EDO for decision. The CRGR operates under a charter that specifically identifies the documents to be reviewed and the analyses, justifications, and findings to be provided. Thus, although the primary responsibility for proper backfit considerations belongs to the organization proposing a communication, the CRGR charter is a key implementing procedure for generic backfitting.

Plant-specific backfits and requests for information are governed by NRC Manual Chapter 0514. In addition, all regional offices and the Office of Nuclear Reactor Regulation have approved procedures that implement the manual chapter. All levels of the NRC staff are responsible for identifying potential backfits, which then are to be handled in accordance with procedures that provide detailed guidance on identification, analyses, justification, and tracking of backfitting items. Training is provided at all staff levels in the principles of management and control of plant-specific backfitting.

Manual Chapter 0514 also provides for licensee claims or appeals regarding plant-specific backfitting determinations. A licensee may claim that an action, which the staff did not consider to be a backfit, is in fact a backfit. In an appeal, a licensee may

- ask that denial of a prior claim of backfit be reversed

- assert that a recognized backfit, which the staff considered to be an adequate protection or compliance exception, does not meet the criteria for the exception

- ask that a proposed backfit, which the staff considered to be a cost-justified substantial safety improvement, be modified or withdrawn

The EDO delegated oversight responsibility of the plant-specific backfitting process to the Director of the Office for Analysis and Evaluation of Operational Data (AEOD). This includes reviewing and concurring with office procedures, conducting training for NRC staff and industry, and informing licensees of program and procedure changes. AEOD conducts an annual assessment of the backfitting process by reviewing plant-specific backfits identified by staff or industry and office procedures and selected records, interviewing office and regional staff, and obtaining industry comments.

In 1989 AEOD conducted the most recent series of NRC staff training sessions at the regional offices and conducted a survey of licensees to determine their perceptions of the backfitting process and obtain specific cost information. In late 1989 and early 1990, the NRC staff conducted a broader survey of licensees concerning the effects of the regulatory process. With regard to backfitting, these surveys indicate that licensees are concerned about the number and overall burden of generic communications, the adequacy of the NRC's consideration of the effects of cost and scheduling, the basis for issuing re-

quirements involving backfits, the NRC's treatment of optional actions and requests as if they were requirements, the negative effects if NRC perceives licensees to be nonresponsive because they do not implement optional actions or because they file backfit claims or appeals, and a need for additional training in backfitting for industry and NRC personnel.

The NRC staff is taking several initiatives to improve the backfitting process and is considering further initiatives. To make backfitting considerations and bases clear to readers, a summary of backfitting considerations was added to generic letters and bulletins beginning in December 1989. This report was prepared to explain the backfitting process to industry and NRC staff. Workshops with industry and NRC staff are planned for the near future. Changes have been proposed to the programs for systematic assessment of licensee performance that would reduce any potential for penalizing licensees for submitting appeals. Senior NRC managers are considering the information gathered from the broad survey of the effects of the regulatory process on licensees to determine what changes may be appropriate. For example, the preliminary report on the broad survey, Draft NUREG-1395, indicates that the staff will examine methods that will take into account the cumulative effects of new requirements.

1 INTRODUCTION

Over the years, issues with regard to what constitutes a backfit and questions on agency policy and practices have been raised inside and outside the agency. This report is intended to address these issues and promote a clearer understanding of the backfit rule and both the generic and plant-specific backfit policies and associated processes that have been adopted by the U.S. Nuclear Regulatory Commission (NRC).

The Commission revised the backfit rule (Title 10 of the *Code of Federal Regulations,* Section 50.109 [10 CFR 50.109]) in 1985 to provide more specific guidance for backfitting decisions and to provide for management control and accountability of backfits. Although the 1985 rule has been superseded, it is included as Appendix A because its statement of considerations provides background information on the development of current practice.

The 1985 rule and the NRC Manual Chapter, which implemented the rule, were vacated by the U.S. Court of Appeals in 1987. The court stated that the rule was ambiguous about whether economic costs would be considered in ensuring or redefining adequate protection of the public health and safety. In 1988, a revised backfit rule was published to clearly state that economic costs cannot be considered (1) when a modification is necessary to bring a facility into compliance with Commission rules or written licensee commitments, (2) when regulatory action is necessary to ensure adequate protection of public health and safety, or (3) when the regulatory action involves defining or redefining the adequate protection standard. The court upheld the 1988 revised rule, which is included as Appendix B.

Backfits are expected to occur as part of the regulatory process to ensure the safety of power reactors. It is important for sound and effective regulation, however, that backfitting be conducted by a controlled and defined process. The NRC backfitting process is intended to provide for a formal, systematic, and disciplined review of new or changed positions before imposing them.

The backfit process enhances regulatory stability by ensuring that changes in regulatory staff positions are justified and suitably defined. For example, even if not needed to meet the standard of adequate protection or to ensure compliance, backfitting is proper if a substantial safety benefit is realized and the costs are justified by the safety benefit.

In its implementing procedures, the Commission has defined two types of backfits, generic and plant-specific. Generic backfits apply to more than one facility while plant-specific backfits apply to only one facility. After management makes appropriate findings, proposed generic backfits are reviewed by the Committee to Review Generic Requirements (CRGR) to determine their compliance with the requirements of the backfit rule and to ensure that new requirements and staff positions contribute effectively and significantly to the health and safety of the public and lead to optimal utilization of NRC and licensee resources. The CRGR Charter (Appendix C to this report) provides specific procedures for handling generic backfits.

Plant-specific backfits are implemented through use of NRC Manual Chapter 0514, "NRC Program for Management of Plant-Specific Backfitting of Nuclear Power Plants," 1988 (Appendix D to this report). This procedure defines the NRC staff responsibilities for implementing the backfit rule for plant-specific applications. The NRC staff, at all levels, is responsible for identifying plant-specific backfits. The cognizant NRC office director or regional administrator determines if the backfit is warranted and the type of analysis or evaluation required, and ensures the proper implementation of the backfitting process.

In late 1989 and early 1990, the NRC conducted a broad survey throughout the industry regarding the effects of NRC regulatory programs on licensees. The results were documented in Draft NUREG-1395, "Industry Perceptions of the Impact of the U.S. Nuclear Power Plant Activities," February 1990. The comments received about backfitting generally confirmed and expanded on concerns that had been expressed in an earlier backfitting survey conducted in April 1989. Senior NRC managers are considering the information received in response to the survey to determine if the NRC should change its regulatory approach. Therefore, this report, which describes the backfitting process as it exists now, could be superseded in some areas by future changes. However, it was considered appropriate to explain the current process and provide documented support for planned training and workshops with industry and NRC staff at this time rather than waiting for ultimate resolution of the issues identified in the survey.

Questions about the backfit process or this report may be addressed to the NRC Office for Analysis and Evaluation of Operational Data (AEOD), which has responsibility for monitoring the backfit process.

2 DISCUSSION

2.1 Nature and Types of Backfits

2.1.1 Background

Backfitting is defined in 10 CFR 50.109 as

(1) the modification of or addition to

- systems, structures, components or design of a facility; or

- the design approval or manufacturing license for a facility; or

- the procedures or organization required to design, construct, or operate a facility;

and

(2) may result from

- a new or amended provision in Commission rules or

- the imposition of a regulatory staff position that is either new or different, from a previously applicable staff position

and

(3) effective after specific dates keyed to the effective date of the backfit rule (see Section 2.1.2 of this report).

Note that the backfit rule and the definition of backfitting apply to cases of compliance and cases of adequate protection as well as to cases of cost-justified substantial safety improvement. They are all backfits, but require different types of justification as discussed further in Section 2.1.3(1) of this report.

The backfit rule applies to nuclear power reactors. The scope of the rule includes all design and hardware aspects of systems, structures, and components as well as supporting activities reflected by procedures and organization.

The rule is intended to encompass only positions or requirements that bring about improvements in safety. Therefore, NRC actions that merely request information and do not impose changes (specifically in hardware, procedures, or organization) are not covered under 10 CFR 50.109, but may be addressed under 10 CFR 50.54(f). The use of 10 CFR 50.54(f) requires an analysis of the burden

to be imposed on responders, but this analysis has a limited scope and depth relative to that required for a 10 CFR 50.109 backfit.

The backfit rule applies to actions that impose positions or requirements on licensees; it does not apply to requested actions that are optional or voluntary. Generally, it does not apply to relaxations.* However, if requirements are reduced but made mandatory, the backfit rule would apply if licensees are required to make the changes in order to achieve a greater level of safety.

The backfit rule does not apply to specific requirements imposed by statute. For example, if a statute requires a revision to license fee schedules, the backfit rule does not apply.

The backfit rule does not apply to purely administrative matters. For example, a change in the number of copies of safety analysis reports that licensees must submit to the NRC would not be covered by the backfit rule.

Different standards apply to the imposition of more stringent safety requirements for standard design certifications (SDCs) or early site permits issued under 10 CFR Part 52. For example, during the pendency of an SDC, backfits of the SDC are permitted only for the sake of compliance or adequate protection. Those standards are not covered in this report.

In its amended (1988) form, the rule requires a backfit analysis, including consideration of associated implementation costs, for all proposed backfits with the following exceptions:

- modifications necessary to bring a facility into compliance with its license or into conformance with written commitments by the licensee

- actions necessary to ensure adequate protection

- actions that involve defining or redefining what constitutes adequate protection

For these exceptions, instead of a backfit analysis, the rule requires a documented evaluation including a statement of the objectives of and the reasons for the backfit and the basis for invoking the exception.

Since 1985, the NRC has issued a number of bulletins, generic letters, and regulatory guides that have been

---

*For generic requirements, the CRGR Charter contains standards for relaxations that do not appear in the backfit rule, as discussed in Section 2.1.3 of this report.

considered backfits.* Many of these actions were exempted from the requirement for a backfit analysis, including cost considerations, because they were considered necessary for adequate protection or compliance. Others were considered to be cost-justified substantial safety enhancements on the basis of a backfit analysis.

2.1.2 Backfit Determination

A backfit involves a modification to the plant, design approval, manufacturing license, procedures, or organization. In addition, (1) a new or revised staff position or requirement must be involved, that is, there must be a change in content or applicability of the previously applicable regulatory staff position (in the direction of increased safety requirements) and (2) this change must be issued after specified dates or milestones.

- **Applicable Regulatory Staff Position**

    A requirement or position already specifically imposed on or committed to by a licensee is called an applicable regulatory staff position. There are several different types of positions, such as

    - legal requirements, as in explicit regulations, orders, and plant licenses and in amendments, conditions, and technical specifications

    - written licensee commitments such as those contained in the final safety analysis report, licensee event reports, and docketed correspondence, including responses to NRC bulletins, generic letters, inspection reports, or notices of violation and confirmatory action letters

    - NRC staff positions that are documented explicit interpretations of more general regulations and are contained in documents such as the Standard Review Plan, branch technical positions, regulatory guides, generic letters, and bulletins

    For the purpose of this report, a change in the applicable regulatory staff position will be subsequently referred to as a new or revised position.

- **Date of Issuance**

    When a new or revised position is issued, it is considered a backfit if it is issued after

    - the issuance of the construction permit for the facility for facilities with construction permits issued after May 1, 1985

    - 6 months before the date of docketing of the operating license application for the facility for facilities with construction permits issued before May 1, 1985

    - the issuance of the operating license for the facility

    - the issuance of the design approval under Appendix M, N or O of 10 CFR Part 50 (now 10 CFR Part 52)

2.1.3 Justification for Imposing Backfits

Section 2.1.3(1) addresses the basic elements of findings, documented evaluations, and backfit analyses required by in the backfit rule. The NRC's internal procedures for implementing the backfit rule address all of these same elements but actually go beyond the rule and contain additional justification requirements as well; these additional requirements are discussed in Sections 2.1.3(2) and 2.1.3(3).

(1) Basic Backfit Justification (Backfit Rule)

    The NRC staff is responsible for identifying plant-specific and generic backfits and for determining if proposed new or revised positions would constitute a backfit. Staff positions are not communicated to licensees unless the NRC official communicating that position determines whether the position is a backfit. At any point during the process, it may be decided to drop the position because further work is not likely to show (a) that the resulting safety benefit is required for compliance or adequate protection or (b) that the action would provide substantial additional overall protection and the direct and indirect costs of implementation would be justified.

    (a) Documented Evaluation (Compliance and Adequate Protection)

        In the case of ensuring compliance with existing requirements or commitments, a backfit analysis is not required. Instead, a documented evaluation of the type discussed in 10 CFR 50.109(a)(6) is prepared and a finding is made that the action is necessary to ensure compliance. The documented evaluation includes a statement of the objectives of and the reasons

---

*As a legal matter, the backfit rule does not strictly apply until the point at which a backfit is required by, for example, a rule or an order. However, for the purpose of this discussion, that legal distinction is not important. The NRC backfit process, including the CRGR Charter and NRC Manual Chapter 0514, is defined on the principle that new positions or requirements are to meet the standards of the rule before they are issued to the licensee(s). New generic positions in documents, such as generic letters, bulletins, and regulatory guides, as well as plant-specific positions, are to be considered and justified as backfits before they are issued. For this reason, they often are discussed in the same way as legally required backfits.

for the action and the basis for invoking the compliance exception.

Similarly, in the case of a backfit needed to ensure adequate protection of public health and safety, a backfit analysis is not required. A documented evaluation of the type discussed in 10 CFR 50.109(a)(6) is prepared and a finding is made that the action is necessary for adequate protection. The documented evaluation includes a statement of the objectives of and the reasons for the backfit and the basis for invoking the adequate protection exception. The concept of what constitutes adequate protection is an evolving standard. It is expected that this standard will continue to change to keep up with new information and with improvements in nuclear power technology. For example, an amendment was recently proposed to 10 CFR 50.61, "Fracture Toughness Requirements for Protection Against Pressurized Thermal Shock Events." This was a case where new knowledge indicated adjustments were needed in the provisions for dealing with vessel embrittlement in order to maintain adequate protection.

For either the compliance case or the adequate protection case, if immediately effective regulatory action is needed, the required documented evaluation may follow the issuance of the regulatory action.

(b) Cost-Justified Substantial Safety Enhancement

For backfits providing a cost-justified substantial safety enhancement, the staff must develop a backfit analysis of the type discussed in 10 CFR 50.109(a)(3) and 10 CFR 50.109(c) and a finding is made that there is a substantial safety benefit to be achieved and that the costs are justified by the benefit. The backfit analysis considers

- how the backfit should be scheduled in light of other ongoing regulatory activities at the facility

- information available concerning any of the following factors as may be appropriate:

  - statement of the specific objective that the proposed backfit is designed to achieve

  - general for description of the activity that would be required by the licensee or applicant in order to complete the backfit

  - potential for change in the risk to the public from the accidental offsite release of radioactive material

  - potential impact of radiological exposure to facility employees

  - installation and continuing costs associated with the backfit, including the cost of facility downtime or the cost of construction delay (i.e., resource burden on licensees)

  - the potential safety impact of changes in plant or operational complexity, including the relationship to proposed and existing regulatory requirements

  - the estimated resource burden on the NRC associated with the proposed backfit and the availability of such resources

  - the potential impact of differences in facility type, design, or age on the relevancy and practicality of the proposed backfit

  - whether the proposed backfit is interim or final and, if interim, the justification for imposing the proposed backfit on an interim basis

For this type of backfit, there first must be a substantial increase in overall protection (or common defense and security), even for requirements that might bring about a net-cost savings. If there is a substantial increase, then the cost justification must be considered. The backfit rule requires the NRC to consider the cost of facility downtime or construction delay as costs associated with the backfit.

Averted *onsite* costs can arise when it is estimated that the backfit will save money for licensees, such as by reducing forced outage rates. These savings are not treated as a benefit (safety enhancement). They are, however, considered as a negative cost, that is, an offset against other licensee costs. Averted *offsite* costs can result from an estimated decrease in accident frequency or severity. These reductions are tied directly to the public health and safety and are considered as a benefit (safety enhancement). "Regulatory Analysis Guidelines of the U.S. Nuclear Regulatory Commission" (NUREG/BR–0058, Rev. 1, May 1984) provides further guidance on this subject.

For this type of backfit, the backfit rule does not require a strict quantitative showing that benefits exceed costs, but rather "that there is a substantial increase in the overall protection of the public health and safety or the common defense and security to be derived from the backfit and that the direct and indirect costs of implementation for that facility are *justified* in view of this increased protection" (emphasis added). Qualitative factors can be considered. Many of the factors to be addressed in the analysis may not be easily quantified and the backfit rule permits consideration of other relevant and material factors.

(2) Regulatory Analyses (Staff Procedures)

Regulatory analyses are generally performed in accordance with the directives and guidance of NUREG/BR–0058 (Rev. 1, May 1984) and NUREG/CR–3568 ("A Handbook of Value–Impact Assessment," December 1983), which describe the need for regulatory analyses and their preparation.* The complexity and comprehensiveness of the analyses should be limited to what is necessary to provide an adequate basis for a decision. NUREG/BR–0058, Section III.A.2, "Scope of the Analysis," states: "The emphasis [in doing the analysis] should be simplicity, flexibility, and common sense, both in terms of the type of information supplied and in the level of detail provided."

For plant-specific backfits, Section 043 of NRC Manual Chapter 0514 requires preparation of regulatory analyses for backfits other than those that fit the adequate protection or compliance exceptions. It also specifies the factors to be included in the regulatory analyses, which include those of a backfit analysis as well as other factors. Thus, this type of regulatory analysis would be the same as a backfit analysis, except that it would contain additional information as well.

For generic backfits, Item IV(b)(5) of the CRGR Charter specifies preparation of regulatory analyses for CRGR review packages. In this case the regulatory analyses may omit some of the factors of a backfit analysis, such as the priority and schedule for implementation, and they may contain additional factors, such as an analysis of alternatives to the pro-

posed action. A typical way of handling this situation for CRGR review packages is to address each backfit analysis factor (which also is specifically listed in the CRGR Charter), making reference to the regulatory analyses if it contains the necessary information. An example of this approach is provided in Appendix E.

Regulatory analyses generally contain a value impact (or cost benefit) analysis; however, as discussed earlier, it would not be appropriate (or permissible) for an adequate protection or compliance backfit to consider the cost in deciding on imposition of the backfit (except for deciding which among several acceptable alternatives to prescribe).

(3) Further Justification (Staff Procedures)

In addition to backfit analyses and regulatory analyses, NRC procedures contain further justification requirements.

For generic backfits, Section IV.B of the CRGR Charter contains a number of other factors to be addressed in all CRGR review packages for new generic requirements or positions. For example, item IV.B(iv) specifies the proposed method of implementation and the concurrence (with any comments) of the Office of the General Counsel. Item IV.B(ix) specifies the necessary findings and standards for relaxations in requirements, which are not addressed in the backfit rule. Finally, Section II.D of the charter exempts compliance and adequate protection cases from the backfit analysis factors and specifies the documented evaluations needed in accordance with the backfit rule.

For plant-specific backfits, Section 043 of NRC Manual Chapter 0514 specifies some of the same additional factors as the CRGR Charter, but only for backfits that are not compliance or adequate protection backfits. The manual chapter further specifies that a proposed plant-specific backfit must be considered for generic backfitting.

2.2 Information Requests

Informal oral information requests are not considered to be backfitting and they should not be used by the staff or accepted by licensees for the purpose of imposing backfits. When written requests cite 10 CFR 50.54(f), requiring a response under oath or affirmation, a statement of the reasons for the request must be prepared and must be approved by the Executive Director for Operations (EDO) or his designee (regional administrators, office directors and their deputies) except when the information is needed to verify compliance with the current licensing basis. As specified in the rule, this is done to ensure that the burden imposed on respondents is justified in view of

---

*It should be noted that the staff is in the process of revising these two guidance documents.

the potential safety significance of the issue to be addressed. As further specified in NRC Manual Chapter 0514 for plant-specific requests, such justification is not needed when seeking information of the type routinely sought for licensing reviews of plants under construction or when there is reason to believe that there is not adequate protection.

Some information requests promulgate new or revised staff positions and request that licensees, in their responses, state whether they will adopt the new positions. Even though these actions do not impose backfits, as a matter of internal staff practice they are identified as backfits and justified accordingly before they are issued, as required by NRC procedures. As discussed in Section 2.1.1, this is often the case with generic letters and bulletins. In the past, backfitting considerations have not been explicitly addressed in the generic letters and bulletins themselves and this has contributed to confusion about whether the actions are backfits. In the future, generic letters and bulletins will contain an explicit statement as to whether the action is considered to be a backfit and, if so, the type of backfit it is considered to be (see Appendix F for a sample).

2.3 Staff Process for Identifying and Imposing Generic Backfits

Backfits that have been identified and justified by the staff and that are intended to apply to one or more classes of commercial nuclear power licensees, first go through the office concurrence chain. The appropriate office director will review the proposed action and disapprove or approve it as a backfit (1) that falls under one of the backfit rule exceptions previously identified or (2) that provides a substantial increase in the overall protection of public health and safety with direct and indirect costs of implementation that are justified in view of this increased protection.

When the office director approves the package, the proposed action and associated justification are forwarded to the Committee to Review Generic Requirements (CRGR) for review. The six-member CRGR normally will discuss the proposal with the sponsoring office to ensure the proposal is well understood, to review its justification, and to make a recommendation to the EDO whether the proposed generic requirement should be issued, issued with modifications, or not issued. If the CRGR recommends disapproval, or recommends major modifications of a proposed requirement, it submits a statement of the reasons for its recommendations to the EDO.

The CRGR was formed in November 1981 and has reviewed the generic requirements or staff positions imposed by the NRC staff since that date. Its charter was revised in 1986 to reflect the 1985 changes to the backfit rule (10 CFR 50.109) and again in 1987 to reflect changes to the NRC organization. The responsibility for supporting CRGR activities and oversight of backfitting was delegated from the EDO to the Director, Office for Analysis and Evaluation of Operational Data (AEOD), in April 1987.

The objectives of the CRGR process are to eliminate or remove any unnecessary burdens placed on licensees, to reduce the exposure of workers to radiation in implementing new requirements, and to ensure the effective use of licensee and NRC resources, while at the same time ensuring the adequate protection of the public health and safety and furthering the review of new, cost-effective generic requirements and staff positions. The committee is chaired by the Director of AEOD and consists of a member each from the Office of the General Counsel (OGC), Nuclear Reactor Regulation (NRR), Research (RES), Nuclear Material Safety and Safeguards (NMSS), and a regional representative. CRGR members are appointed by the EDO (the NRC General Counsel concurs in the appointment of the OGC member).

The types of documents to be considered by the CRGR include

- staff papers proposing the adoption of rules or policy statements affecting power reactors

- staff papers proposing new or revised rules including advanced notices

- proposed new or revised regulatory guides, Standard Review Plan (SRP, NUREG–0800) sections, and branch technical positions

- proposed generic letters, multi-plant orders, show cause orders, and generic information requests under 10 CFR 50.54(f)

- proposed bulletins and unresolved safety issue NUREGs

- new or revised standard technical specifications

- any generic correspondence to licensees that may reflect or interpret new NRC staff positions

Evaluations and approvals of generic topical reports are examples of documents that sometimes need review. A large majority of these documents do not contain any new requirements or positions; however, some of them do and they are reviewed by the CRGR.

Examples of approved requirements that do not require CRGR review are (1) positions or interpretations contained in the above documents that were issued before November 12, 1981 or (2) positions taken after

November 12, 1981, that already have been approved through the established generic review process.

In reviewing proposed new staff positions or requirements, the committee specifically focuses on (1) the need for a new requirement and whether it may have any adverse effect on safety and (2) if not required for adequate protection or compliance, whether the new requirement provides a substantial improvement in safety and is cost-justified. In conducting this review, the CRGR normally will consider the factors specified for backfitting as discussed in Section 2.1.3(1) as well as additional factors as discussed in Sections 2.1.3(2) and 2.1.3(3).

For those rare instances where it is judged that an emergency action is needed to protect the health and safety of the public, no prior review by the CRGR is necessary. However, the CRGR Chairman is notified by the office originating the action. The objective of and reason for the emergency action requirements are documented and reported to the committee for information and are included in a report to the Commission.

For each proposed requirement not requiring emergency action, the proposing office identifies the requirement as either Category 1 or 2. Category 1 requirements are those that the proposing office rates as urgent and are approved or otherwise dealt with within two working days of receipt by the CRGR. Category 2 requirements are those that do not meet the criterion for designation as Category 1. These are scrutinized carefully by the CRGR on the basis of oral discussion and written justification. Such justification is submitted by the proposing office along with the proposed requirements in advance of CRGR discussions. Meetings are generally held at regular intervals and agendas are issued by the CRGR Chairman one to two weeks in advance of each meeting, except for Category 1 items. Available background material on each item to be considered by the committee is issued to each CRGR member as it is received to permit sufficient advance review.

Fifteen copies of each review package are submitted to CRGR. The following type of information is submitted (see the CRGR Charter, Appendix C, for specific details):

- the proposed generic requirement or staff position

- supporting documents

- the proposed method of implementation

- a backfit analysis for cost-justified enhancements, generally conforming to the directives and guidance of NUREG/BR-0058 and NUREG/CR-3568

- category of reactor to which the generic requirement or staff position is to apply

- the office director's determinations

The CRGR may recommend approval, revision, or disapproval or that further work be done by the staff and/or public comment be sought.

A written response is required from the cognizant office to report agreement or disagreement with the CRGR recommendations documented in CRGR meeting minutes.

The CRGR staff ensures that there is an archival system for keeping records of all packages submitted, actions by the staff, summary minutes of CRGR consideration of each package, including corrections and recommendations by the committee. The submitted packages and the summary minutes for a meeting are released to the Public Document Room after the NRC has taken action on the matters discussed (e.g., issuance of a generic letter or bulletin) or after the Commission has considered the matters in a public forum (e.g., public meeting on a proposed rule).

The CRGR staff prepares a report that is submitted by the EDO to the Commission each month. The report provides a brief summary of CRGR activities. The report is distributed as an enclosure to the EDO Weekly Highlights.

Figure 1 provides a schematic representation of how new generic requirements and staff positions are developed, revised, and implemented.

2.4 Staff Process for Identifying and Imposing Plant-Specific Backfits

As noted previously, plant-specific backfitting involves positions unique to a particular plant, whereas generic backfitting involves the imposition of the same or similar positions on more than one plant. To be a plant-specific backfit, the requirement or position will involve (1) only one plant (sometimes including identical units at one site), (2) a new or revised requirement or staff position, and (3) a schedule for the imposition after key dates specified in the backfit rule.

It is important that the necessity for making backfit determinations not inhibit the normal informal dialogue between NRC staff (e.g., technical reviewers and inspectors) and the licensee. The intent is to manage backfit imposition and not to constrain or eliminate suggestions or inquiries in areas within the scope of 10 CFR 50.109. Only when these conversations convey a staff position that a licensee must change the design, construction, or

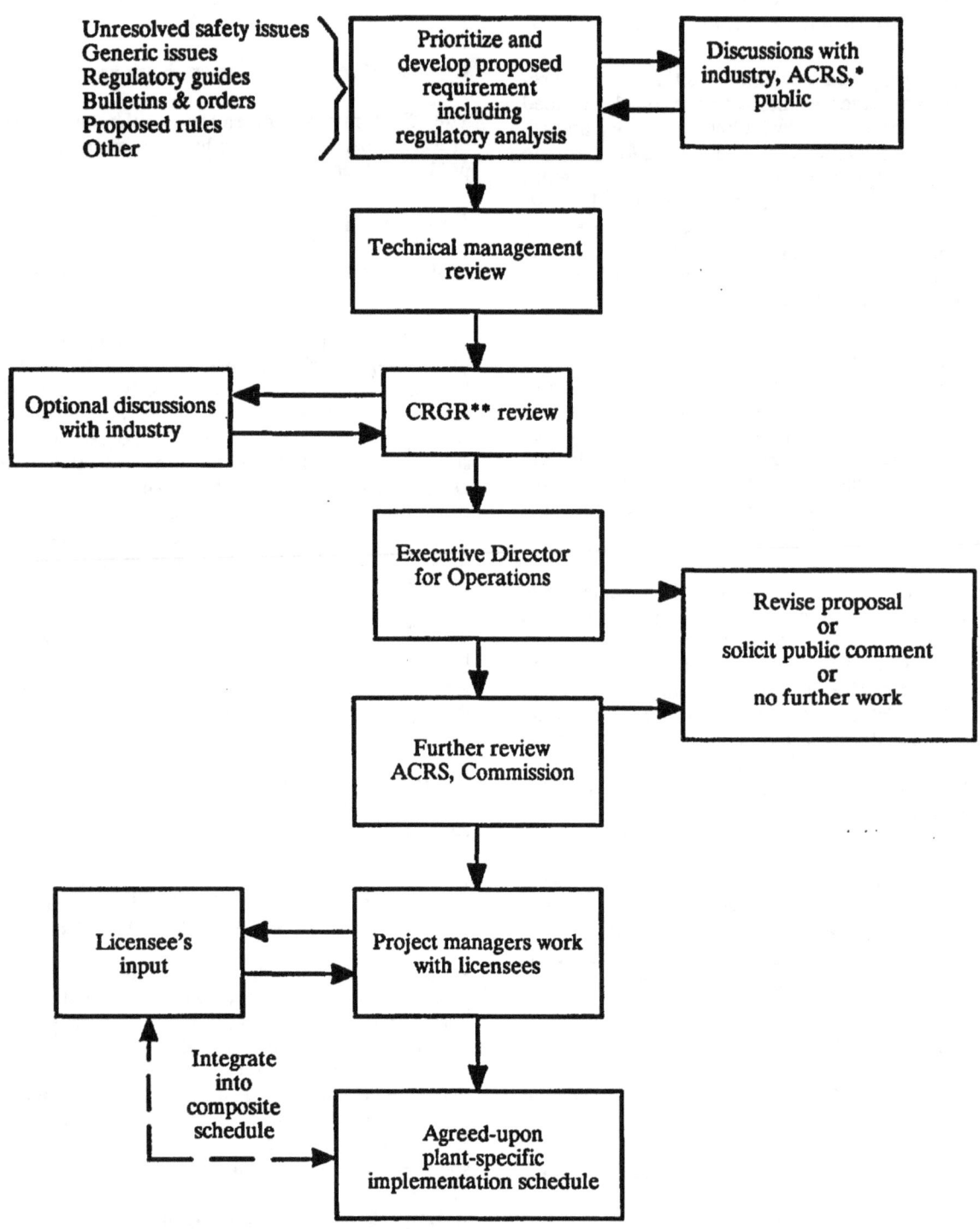

Unresolved safety issues
Generic issues
Regulatory guides
Bulletins & orders
Proposed rules
Other

Prioritize and
develop proposed
requirement
including
regulatory analysis

Discussions with
industry, ACRS,*
public

Technical management
review

Optional discussions
with industry

CRGR** review

Executive Director
for Operations

Revise proposal
or
solicit public comment
or
no further work

Further review
ACRS, Commission

Licensee's
input

Project managers work
with licensees

Integrate
into
composite
schedule

Agreed-upon
plant-specific
implementation schedule

*Advisory Committee on Reactor Safeguards
**Committee To Review Generic Requirements

Figure 1 Schematic representation of new requirements review

operation of a facility would a backfit determination be needed.

In this context, it should be noted that actions proposed by the licensee are not backfits, even though such actions may result from normal discussions between the staff and the licensee concerning an issue and even though the change or additions proposed by the licensee may otherwise meet the definition of a backfit.

The imposition of plant-specific backfits is governed by NRC Manual Chapter 0514, which establishes the staff requirements and guidance for implementation for this aspect of the backfit rule. The primary objective of the manual chapter is to ensure that plant-specific backfits are required only (1) if necessary to provide an adequate level of safety, (2) if necessary to ensure compliance with Commission rules, orders, or written licensee commitments, or (3) to provide a cost-justified safety enhancement after approval of the required backfit analysis. If no backfit analysis is required, the appropriate office director or regional administrator is to provide a documented evaluation that provides the basis for invoking one of the exceptions.

The manual chapter identifies the NRC staff members responsible for implementing the procedure and assuring that process controls are in place. For example, staff members at all levels are responsible for identifying potential plant-specific backfits. The office directors for NRR and NMSS and the regional administrators are responsible for the final decision on whether a backfit is required and, if so, to approve the backfit analysis or the documented evaluation. Further, each office is required to have a specific office procedure providing guidance in the identification, handling, imposition, and tracking of plant-specific backfits.

Following approval of the regulatory analysis (backfit analysis) or documented evaluation by the appropriate office director or regional administrator, review (if any) by the EDO, and issuance of the backfit requirement to the licensee, the licensee may implement the backfit or appeal it. Following an appeal and subsequent final decision by the appropriate office director or EDO, if the appeal has been denied the licensee will normally implement the backfit. If the licensee still does not elect to implement the backfit, it may be imposed by order of the appropriate office director.

Implementation of a plant-specific backfit is normally accomplished on a schedule negotiated between the licensee and NRC. Scheduling criteria include the importance of the backfit relative to other safety-related activities under way, such as the plant construction or maintenance planned for the facility in order to maintain a high-quality of construction or operations. For plants that have integrated schedules, the integrated scheduling process is used for this purpose.

A staff-proposed backfit may be imposed by order before completing any of these procedures, if the NRC official who authorizes the order determines that immediate imposition is necessary to ensure public health and safety or the common defense and security. In such cases, the EDO shall be notified promptly of the action and a documented evaluation prepared (if possible in time to be issued with the order).

If immediate imposition is not necessary, staff-proposed backfits should not be imposed, and plant construction, licensing action, or operation should not be interrupted or delayed by NRC actions during the staff's evaluation and backfit transmittal process, or a subsequent appeal process, until final action is completed.

The proposing headquarters office or regional office manages each proposed plant-specific backfit using the NRC plant-specific backfit tracking system. This system provides references to all documents issued or received by NRC staff relative to plant-specific backfits, including requests, positions, statements, and summary reports. Specific details on this system are found in office implementing procedures.

As stated earlier, the EDO has delegated responsibility for oversight of the plant-specific backfitting process to the Director of AEOD. This includes reviewing and concurring with office procedures, conducting training for NRC staff and industry, and informing licensees of program and procedure changes. An annual assessment is conducted that includes review of plant-specific backfits identified by staff or industry, review of office procedures and selected records, interviews with office and regional staff, and obtaining industry comments on the backfitting process.

A graphical description of the plant-specific backfit process is given in Figure 2.

2.5 Filing a Backfit Claim or Appeal

A proposed staff position not identified by the NRC staff as a backfit may be claimed to be a backfit by a licensee. All licensee claims are to be sent in writing to the office director or regional administrator of the NRC employee who issued the position with a copy to the EDO. A licensee claim that a requested action is a backfit needs to be promptly addressed and evaluated to determine whether it is, in fact, a backfit. A report to the EDO and a response to the licensee should be forwarded within three weeks after receipt of the claim indicating the results of the determination and the plan for resolving the issue.

Appeals with regard to backfit determinations are generally of two types and involve two different situations:

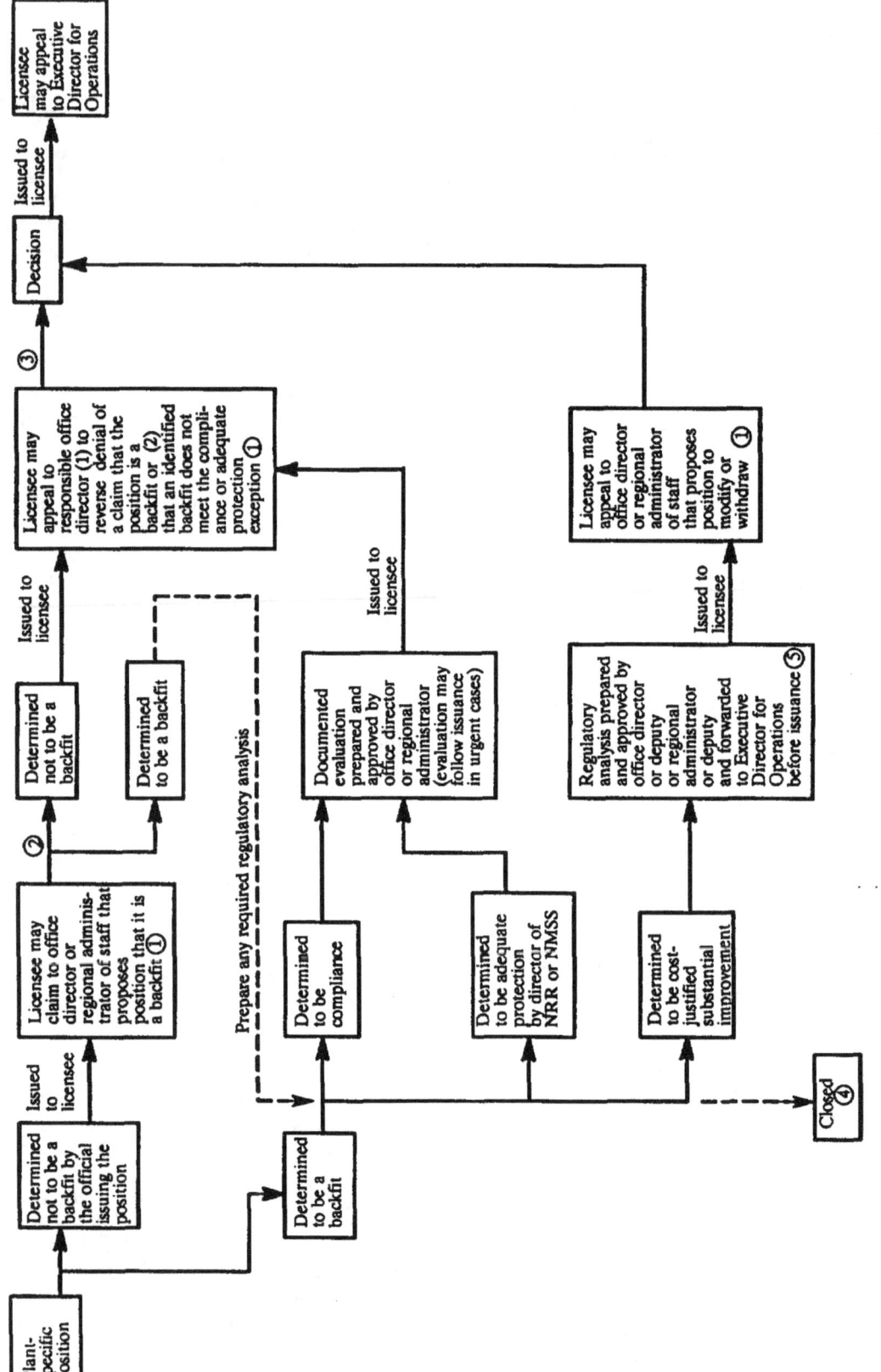

Figure 2 NRC plant-specific backfit, claim and appeal process defined in NRC Manual Chapter 0514

Notes:
1. Copy to be addressed to the Executive Director for Operations.
2. Report to Executive Director for Operations and inform licensee within 3 weeks of results of determination and plans to resolve issue.
3. Report to Executive Director for Operations within 3 weeks on plan for resolving and inform licensee promptly and periodically of plans.
4. Staff may decide the backfit is not likely to be justified and close the action.
5. Approval by Executive Director for Operations before issuance is not required.

(1) appeal to an office director or regional administrator proposing to modify or withdraw a backfit for which a regulatory analysis has been prepared and transmitted to the licensee

(2) appeal to the responsible program office director (a) to reverse a denial of a prior licensee claim that an action is a backfit or (b) to determine that a backfit that the staff found to meet the adequate protection exception or the compliance exception does not meet the exception

Licensees should address an appeal of a proposed backfit to the appropriate office director or regional administrator with a copy to the EDO. The appeal should indicate deficiencies in the staff's analysis or provide other information in support of the appeal. In all cases, the appeal should include sufficient documentation to justify the position taken. The office director or regional administrator, within three weeks, reports to the EDO on the plan for resolving the issue and informs the licensee in writing of the staff plan. Appropriate documents should be included in the backfit tracking system. Licensees shall not be penalized by the staff for raising backfit questions or filing backfit appeals. As stated in NRC Manual Chapter 0514, if immediate imposition is not necessary, staff proposed backfits should not be imposed and plant construction, licensing action, or operation should not be delayed during an appeal.

The decision of the office director or regional administrator on a plant-specific backfit appeal may be appealed to the EDO, in which case the EDO will resolve the appeal and state the basis.

Summaries of all appeal meetings are prepared promptly, provided to the licensee, and placed in appropriate public document rooms. After an appeal and subsequent final decision by the appropriate office director or regional administrator or the EDO, if the appeal has been denied, the licensee may implement the backfit resulting from the decision. If the licensee does not elect to implement the backfit, it may be imposed by order of the appropriate office director. Additional details on the backfit claim and appeal process can be found in NRC Manual Chapter 0514. A graphical description of this process is given in Figure 2.

The regional offices and the Office of Nuclear Reactor Regulation have procedures that govern the specific methods of reviewing backfit appeals. These procedures provide further detail beyond the requirements of NRC Manual Chapter 0514 and the details differ somewhat among the offices. However, all of the procedures conform to the provisions of Manual Chapter 0514 discussed above

2.6 Current Status

In 1989 the Office of AEOD conducted the most recent series of NRC staff training sessions at the regional offices. AEOD also, in April 1989, conducted a survey of licensees to determine their perceptions of the backfitting process and obtain specific cost information. In addition, in late 1989 and early 1990, the NRC staff conducted a broader survey with regard to the effects of the regulatory process on licensees. With regard to backfitting, licensees expressed concern about the number and overall burden of generic communications, the adequacy of NRC's consideration of cost and schedule impacts, the basis for issuing requirements involving backfits, the treatment of optional actions and requests as if they were requirements, the negative effects if NRC perceives licensees to be unresponsive because they do not implement optional actions or because they file backfit claims or appeals, and a need for additional training in backfitting for industry and NRC personnel.

The NRC staff is taking several initiatives to improve the backfitting process and is considering further initiatives. To make backfitting considerations and bases clear to readers, a summary of backfitting considerations was added to generic letters and bulletins beginning in December 1989. This report has been prepared to explain the backfitting process to industry and NRC staff. A series of workshops with industry and NRC staff is planned for the near future. Changes have been proposed to the program for systematic assessment of licensee performance that would reduce any potential for penalizing licensees for submitting appeals. Senior NRC managers are considering the information gathered from the broad survey of the effects of the regulatory process on licensees to determine what changes may be appropriate. For example, the preliminary report (Draft NUREG-1395) indicates that the staff will examine methods to take into account the cumulative effects of new requirements.

3 QUESTIONS AND ANSWERS ON BACKFITTING

During the conduct of staff training and in the communication with licensees, a number of questions and observations have been raised regarding the NRC policies and practices for backfitting. These questions and observations have been grouped into six general categories and are presented below followed by the approved staff response.

3.1 Backfit Determination and Imposition

(1) *A number of generic letters and bulletins recently issued request or require actions, yet there is no backfit analysis accompanying the documents. Is this appropriate?*

Many of the bulletins and generic letters issued in 1988 and 1989 were not justified by a backfit analysis simply because they were determined to fall under the compliance exception listed in 10 CFR 50.109. When action is needed to ensure compliance with existing regulations or to ensure that an adequate level of protection is maintained, a backfit analysis is not required. However, a documented evaluation is needed to support the use of the exception.*

The backfit analyses or documented evaluations are available in the Public Document Room. However, this was not readily apparent in the past because they were not cited in the generic letters and bulletins themselves. In the future, backfit analyses or documented evaluations will be cited in the generic communications. An example is provided in Appendix F to this report.

(2) *Where does a confirmatory action letter fall in the backfit process?*

A confirmatory action letter (CAL) is issued to confirm a licensee's agreement to implement specific actions, which can include agency requirements and staff positions. The CAL merely documents a licensee's agreement and does not impose or implement any new or revised staff positions or requirements. Thus, it falls outside of the backfit process because the licensee has volunteered to take the course of action identified in the letter.

(3) *The backfit rule does not provide a clear explanation of the criteria the NRC will use to document the need for a new "level of protection."*

Specific criteria have not been established to determine when it may be necessary to redefine adequate protection. See *Federal Register*, Vol. 53, No. 108, (June 6, 1988), pages 20608 (col. 3) and 20609 (col. 1) (Appendix B to this report).

(4) *How does the backfit rule apply to new staff positions that reflect an evolving understanding of technical issues?*

New or revised staff positions are backfits when they are imposed on licensees and result in a change in structures, systems, design, or procedures (as described in 10 CFR 50.109). A backfit analysis is required whenever new or revised positions are imposed to achieve cost-justified substantial safety enhancements. A backfit analysis is not required if the new or changed position is imposed to bring a facility into compliance or if it is necessary to provide assurance of adequate protection. In those cases, however, a written evaluation is needed to provide the objectives of and reasons for the modification and the basis for invoking the exception.

An evolving understanding of issues does not, by itself, define which category fits a particular backfit. Judgment must be applied to the facts of each particular case to determine whether the backfit is for compliance, to provide adequate protection, to redefine adequate protection, or to achieve a cost-justified substantial safety enhancement. For example, with regard to compliance, the 1985 statement of considerations for 10 CFR 50.109 indicates that "the compliance exception is intended to address situations where the licensee has failed to meet known and established standards of the Commission because of omission or mistake of fact....new or modified interpretations of what constitutes compliance would not fall within the exception...."

(5) *Must backfits be identified by the staff when they are imposed?*

Yes, plant-specific and generic backfits must be determined in advance and the proper procedures followed before imposition. For example, a backfit analysis is required for actions that are cost-justified substantial safety enhancements. If a new requirement or staff position meets the compliance exception or the adequate protection exception, a backfit analysis is not required, but the action is still

---

*As was stated earlier, generic letters and bulletins do not impose backfits. Therefore, they are not required by regulation to be accompanied by backfit analyses or documented evaluations. Nonetheless, it is NRC practice to justify them as backfits, if appropriate, before issuing them.

considered as a backfit and a documented evaluation providing the objectives of and the reasons for the modification and the basis for invoking the exception is needed.

(6) *Is a backfit analysis needed for information requests that are verbally communicated? Many such requests can represent a significant burden to a licensee.*

Oral information requests do not fall under the backfit rule. They should not be used by the staff or accepted by licensees for the purpose of imposing backfits.

NRC management should be informed of inappropriate requests for information in order to assure proper authorization and justification. NRC supervisors and managers are sensitive to this issue and licensees should not be penalized by the staff for raising it.

(7) *Is it appropriate for the NRC staff to rely on informal or formal communications to other licensees as official NRC positions? What about NRC tacit approval of documents?*

Informal or formal communications to one licensee are not official positions to all licensees. Section 053 of Manual Chapter 0514 identifies what can be applied as official staff positions in a plant-specific context. They are legal requirements such as contained in explicit regulations, orders, and plant licenses; written commitments such as contained in final safety analysis reports, licenses event reports, and docketed correspondence; and documented, approved explicit interpretations such as contained in the SRP, branch technical positions, regulatory guides, generic letters, and bulletins. Orders, licenses, and written commitments are applicable only to a particular licensee.

If the staff previously exempted a licensee from a legal requirement or approved position, it is not applicable to that licensee for the purpose of backfit consideration. Explicit exemption would be done formally in writing. The Appendix to NRC Manual Chapter 0514 discusses tacit approval under reanalysis of issues. Two situations are covered. In the first case, staff review of a previously accepted licensee action or program may result in a requested change. This would be classified as a backfit because it represents a change in a previous staff position and would require a backfit analysis (or a documented evaluation if it meets one of the exceptions listed in the backfit rule). In the second case, a licensee submittal committing to a specific course of action that has not received timely NRC staff review is implemented by the licensee. In this case, it is considered

that the NRC staff tacitly accepted the licensee's action since timely notice to the contrary was not given. If the NRC staff subsequently adopts a different position and requests a change in the licensee action, this change may be classified as a backfit and thus require a backfit analysis (or a documented evaluation if it meets one of the exceptions listed in the backfit rule).

(8) *Is it a plant-specific backfit to apply an approved and issued regulatory guide to an operating plant?*

As part of the generic review process, the responsible office director determines and the CRGR reviews which plants or groups of plants are affected by new or modified regulatory guide provisions. Implementation in accordance with the generic applicability is not an additional plant-specific backfit and is, therefore, not governed by the plant-specific backfit procedures. A licensee may appeal, however, and assert that the generic analysis does not justify the backfit.

Any staff-proposed plant-specific implementation of a regulatory guide provision, whether orally or in writing, for a plant not encompassed by the generic implementation determination is, however, considered a plant-specific backfit. In other words, staff action with regard to a specific licensee that expands on, adds to, or modifies a generically approved regulatory guide, such that the position taken is different than intended in the generic positions, is a plant-specific backfit.

(9) *How is a compliance backfit affected if a licensee formally withdraws or substantially revises the commitment that forms the basis for the compliance backfit?*

Licensees are free to change commitments that have not been imposed by rule or order. However, doing so may raise questions about staff acceptance of licensee programs. Of course, if the revised commitment is fully acceptable to the staff, then the superseded or outdated commitment would not be used as a basis for a compliance backfit. If, on the other hand, the revised commitment is not acceptable, then the previous commitment or its equivalent may form the basis for a compliance backfit. Circumstances and judgment would play a significant role in this case. For example, the date of the original commitment and that for the revised commitment could be important. It would not be appropriate for the NRC staff to cite a previous commitment that had been revised with staff knowledge and tacit approval for several years.

(10) *How does one appeal a generic backfit?*

Licensees may certainly appeal generic backfits as they may appeal any staff position. However, the

CRGR Charter, which is the NRC's generic back-fitting procedure, does not address appeals. Thus, in appealing generic backfits, the staff advises licensees to follow the guidelines in NRC Manual Chapter 0514 for appealing plant-specific backfits to the extent practical. (In the recent past the EDO has referred such appeals to the CRGR to obtain its recommendations before making a decision.)

3.2 Generic Backfits

(1) *Does CRGR look at the difference between generic letters and bulletins?*

Yes, when CRGR reviews generic communications consideration is given to the form and its application. The general guidelines used by the CRGR are that bulletins should be used to request action by licensees on a short-term basis to correct or address a safety concern for which timely action is necessary and that generic letters should be used to request information from licensees and to transmit information regarding a new staff position. In general, generic letters are used to clarify NRC policy on how the agency intends to implement a regulatory requirement, assist the agency in determining whether new requirements are needed, or seek information on licensees conformance to existing staff positions.

(2) *Why is 10 CFR 50.54(f) cited in many generic letters?*

When 10 CFR 50.54(f) is cited in a generic letter, it is to establish a basis for requiring a response. This may be to determine whether the agency will take action regarding the specific license or it may be done to determine whether a generic requirement is needed based on the information obtained.

(3) *It seems like the staff too frequently claims that a new staff position is consistent with existing Commission rules and positions just to avoid performing a backfit analysis.*

The compliance exception in 10 CFR 50.109 is certainly meant to be used only when specifically authorized and justified by the appropriate office director. Further, the CRGR Charter provides added assurance that new requirements and staff positions are fully consistent with the provisions of the backfit rule. In reviewing proposed generic requirements that are identified as compliance issues, the committee considers whether they are needed to ensure compliance with existing requirements or whether they represent a new staff position that needs to be reviewed under other provisions of 10 CFR 50.109.

(4) *It has been our observation that the staff estimate of installation and continuing costs associated with the backfit is often grossly underestimated.*

During its review, CRGR takes into account staff estimates of the required licensee and staff resources to implement the requested action. Frequently staff estimates are given in terms of a range as well as an average cost because it is recognized that some licensees may be required to spend more than other licensees. The CRGR review usually focuses on what is projected to be an average cost over all affected facilities. For adequate protection and compliance actions, cost estimates provided to CRGR have been used for background information and not as a condition for acceptance.

As part of the AEOD April 1989 survey, a questionnaire requesting estimates with actual costs was included. The purpose for requesting this information was to assist the staff and CRGR in evaluating estimated costs for proposed actions. A preliminary review of the licensee's responses indicate that, for the most part, the staff estimates have been reasonable and have not been grossly underestimated in relation to average costs.

(5) *How does the agency ensure that the backfit rule is properly implemented in issuing generic communications such as bulletins and generic letters?*

As discussed in the responses to other questions, before any generic communications such as bulletins and generic letters are issued, the proposed staff positions are reviewed for the method and impact of implementation by the responsible office director and, in turn, by the CRGR. A specific focus of the CRGR review is the basis for the generic communication and whether backfit considerations have been appropriately addressed by the staff.

(6) *It seems that the staff is circumventing the requirements of 10 CFR 50.109 by citing 10 CFR 50.54(f) as the basis for imposing major new regulatory requirements.*

When 10 CFR 50.54(f) is cited in a generic letter or bulletin, it simply establishes a requirement to submit a response to the letter. Thus, there is no intent in citing 10 CFR 50.54(f) to circumvent 10 CFR 50.109. To the contrary, although generic letters and bulletins do not impose new or revised staff positions, they are reviewed by the CRGR to ensure that the provisions of the backfit rule are implemented. In the future, generic letters and bulletins will cite the backfit analysis or other evaluation performed in this regard.

(7) *The spirit and intent of the backfit rule does not appear to have been met in all cases, as an example, issuance of*

Bulletin 88–11 is completely lacking any 10 CFR 50.109 justification.

Although the justification was not printed in the bulletin, NRC Bulletin 88–11, "Pressurizer Surge Line Thermal Stratification," was justified as a backfit. It is an example of a backfit that was determined by the responsible NRC official to be required as a matter of compliance with existing requirements and commitments. The CRGR reviewed the bulletin and concurred. The regulations currently require licensees to meet the applicable codes of the American Society of Mechanical Engineers (ASME), *Boiler and Pressure Vessel Code.* Because of the staff's concern with the integrity of the surge line, licensees were requested to perform their fatigue analysis in accordance with the latest ASME Section III requirements that incorporate high cycle fatigue analysis. The justification provided by the staff was that previously unconsidered thermal stratification phenomenon may invalidate the existing analysis performed to confirm the integrity of the surge line.

Subsequently, it was understood that some licensees believed that the staff's rationale was in error because they were not committed to the latest ASME Section III requirements by virtue of their license commitment. However, the issue became moot because these licensees undertook the analysis voluntarily in view of the safety importance of the issue and the fact that previous versions of the ASME Code did not completely address the concern.

3.3 Plant-Specific Backfits

(1) *If an inspector has previously accepted (i.e., provided tacit approval of) a licensee's method, does a specific request for change constitute a backfit and if so, is a backfit analysis required?*

A new or revised staff position affecting the design of systems, structures, and components or the procedures or organization required to design and construct or operate a facility after issuance of the operating license is a backfit. Whether a backfit analysis is required depends on the basis for the backfit. A backfit analysis is required when the backfit would result in a cost-justified substantial safety enhancement. If a determination is made that the action is needed to provide an adequate level of protection or required to bring a facility into compliance, then no backfit analysis is required. In these cases, a documented evaluation of lesser scope is needed as discussed in the response to previous questions.

Cases where an inspector provides tacit approval are relatively rare. Simply not challenging a licensee's

practice normally would not be considered tacit approval. The only example provided in Manual Chapter 0514 is a case where the NRC has indicated tacit approval by not acting in a reasonable time on a licensee submittal and the licensee has moved ahead to implement the proposal described in the submittal. For the purpose of this question, it would most likely arise in connection with review of a licensee response to an inspection report.

Explicit approval could be provided in an inspection report that states that a particular approach is acceptable. However, conclusions of that nature are usually made in safety evaluation reports rather than inspection reports.

(2) *What is the definition of "timely" in the context of approval?*

The appendix to Manual Chapter 0514 provides the following:

- when the licensee has made a submittal committing to a specific course of action to meet an applicable position

- the licensee has moved ahead in the intervening time to implement the proposed action

- the staff did not provide a response for an extended period (or within a reasonable time not delaying the applicants implementation plans

then subsequent staff action to make changes is (or may be considered) a backfit.

There is no specific time period assigned. Some submittals may require a detailed analysis and could be expected to take several months to complete, while others are administrative and can be completed in several weeks. Discussions need to be held with licensees relating to the agency's progress in reviewing submittals in order to reduce the probability of misunderstandings, excessive delays, and the need for backfit determinations.

(3) *Is the guidance contained in the NRC Inspection Manual approved positions?*

No, inspection procedures are not approved staff positions, which is the reason they are not reviewed by CRGR. They exist only for staff use in conducting inspections. NRC inspection procedures govern the scope and depth of staff inspections associated with licensee activities, such as design, construction, and operations. They define those items the staff is to consider in its determination of whether the licensee is conducting its activities in a safe manner.

Licensees cannot be required to implement positions discussed in an inspection procedure or manual unless the same positions exist in the form of an approved regulatory staff position. Examples of approved staff positions are described in Manual Chapter 0514 and include the SRP, branch technical positions, regulatory guides, generic letters, and bulletins.

(4) *Are NRR office letters considered approved staff positions?*

No, office letters issued by the Office of Nuclear Reactor Regulation (NRR) fall into the same category as inspection procedures and do not constitute approved staff positions. They are not reviewed by CRGR and exist solely as guidance for staff within NRR.

(5) *Are staff positions that reject an industry practice that was previously approved (either tacitly or explicitly) considered to be backfits?*

A change in staff position regarding previously approved industry practice would be considered a backfit as described in the appendix to Manual Chapter 0514. A backfit analysis is required (or a documented evaluation is required if the action meets the exceptions listed in 10 CFR 50.109). If the revised staff position had generic implications, CRGR review would be needed.

Tacit approval is not broadly defined. The only example given in Manual Chapter 0514 is where the NRC has indicated tacit approval by not acting in a reasonable time on a licensee submittal and the licensee has moved ahead to implement the proposed action described in the submittal.

(6) *The distinction between a staff recommendation that asks a licensee to consider a proposed action and one that directs the licensee to take a proposed action is sometimes difficult to determine. How does the backfit process address this fine but important difference?*

In the conduct of agency business, there are many occasions when the staff will suggest or even recommend that licensees consider various actions. Such suggestions and recommendations are not necessarily backfits and licensees may evaluate such recommendations and make the appropriate decision on implementation.

Discussion or comments by the NRC staff identifying deficiencies or weaknesses, whether in meetings or written reports, do not constitute backfits. Definitive statements to the licensee directing a specific action to satisfy staff positions are backfits unless the action is consistent with an explicit regulatory staff position applicable to that facility (see the response to Question 3.1(8) for further discussion). In a similar manner, pressure upon a licensee to adopt a specific staff position (for example to have a program found acceptable) would be prohibited unless the action is consistent with an explicit staff position applicable to that facility.

(7) *There seems to be an apparent trend to inspect licensees to a rising standard of acceptability without an attendant modification to the specific regulatory requirements. How does the backfit process ensure that the inspection standards are properly controlled?*

Inspectors are expected to look beyond mere compliance with regulations and to focus on the safety implications and margins at each facility. As a result, licensees may be encouraged to consider program enhancements and other actions. Licensees are expected to evaluate such suggestions and recommendations and make a decision on implementation; however, such informal requests are not requirements or staff positions. Further, such suggestions or recommendations are not within the scope of the backfit process. The staff should be questioned regarding the safety significance, authority, or justification of any recommendations whenever the basis is not clear. Licensees shall not be penalized by the staff for such questioning.

(8) *How far can an inspector go in interpreting NRC rules in developing inspection findings and requiring licensee actions without performing a backfit analysis?*

In the normal course of inspecting to determine whether the licensee's activities are being conducted safely, inspectors may examine and make findings in specific technical areas where prior NRC positions and licensee commitments do not exist. Examining such areas and making findings are not considered backfits. Likewise, discussion of findings with the licensees is not considered a backfit. If during such discussions, the licensee agrees that it is appropriate to take action in response to the inspector's findings, such action is not a backfit provided the inspector does not indicate that the specific actions are the only way to take corrective actions. On the other hand, if the inspector indicates that a specific action must be taken, such action is a backfit unless it is consistent with an applicable regulatory staff position (see the response to Question 3.1(8) for further discussion). Further, if the licensee provides a written claim that the inspector's findings are a backfit, the staff must make a specific backfit determination. Examples can be found in the appendix to Manual Chapter 0514 (see Appendix D to this report).

(9) *What are the ground rules for applying the Standard Review Plan in operating license reviews? Can the staff*

modify the acceptance criteria for specific cases or when a safety concern has been identified?

The SRP delineates the scope and depth of staff review of licensee submittals associated with various licensing activities. It is an NRC staff interpretation of measures which, if taken, will satisfy the requirements of the more generally stated, legally binding body of regulations primarily found in 10 CFR. Since October 1981, changes to the SRP are reviewed and approved through a generic review process involving the CRGR and the extent to which the changes apply to classes of plants is defined. Consequently, application of a current SRP in a specific operating license review is not, in general, a plant-specific backfit, provided the SRP was effective six months before the start of the operating license review. Asking questions of an applicant for an operating license to clarify staff understanding of proposed actions to determine whether the actions will meet the intent of the SRP is not considered a backfit.

On the other hand, using acceptance criteria more stringent than those contained in the SRP or taking positions more stringent than or in addition to those specified in the SRP, whether in writing or orally, is a plant-specific backfit. During meetings with the licensee, staff discussion or comments regarding issues and licensee actions volunteered that are in excess of the criteria in the SRP usually do not constitute plant-specific backfits; however, if the staff indicates that a specific action in excess of the already applicable staff position is the only way for the staff to be satisfied, the action is considered a plant-specific backfit whether or not the licensee agrees to take such action. It should also be recognized, however, that a verbally implied or suggested action should not be accepted by a licensee as an NRC position of any kind, backfit or not; only written and authoritatively approved position statements should be taken as NRC positions.

(10) *Is it appropriate for the staff to use the latest version of the SRP in the review of license amendment requests and other changes?*

There is not a single answer. In the review of a license amendment request, the staff should consider the guidance in the implementation section of the SRP and in Manual Chapter 0514 and exercise judgment in determining the applicability of current SRP.

During reload or other reviews subsequent to issuance of the operating license, staff-proposed positions with regard to technical matters not related to the changes proposed by a licensee are considered to be backfits.

(11) *Assume that a plant has not complied with an approved staff position that is committed to in the FSAR and the staff's safety evaluation report (SER) was written on the basis that the staff position would be implemented. However, the position has not been implemented. Is it a backfit to impose the position on the licensee after issuance of the operating licensee?*

Generally, it does not appear that the staff would be changing its position in this case. If there is no change of positions, imposing the licensee's commitment would not be a backfit.

(12) *Do plant-specific orders come within the scope of the backfit rule? What about those that confirm licensee actions?*

An order issued to cause a licensee to take actions that are not otherwise applicable regulatory staff positions is a plant-specific backfit. An order effecting prompt imposition of a backfit may be issued before completing any of the backfit procedures, if the appropriate office director determines that prompt imposition is necessary.

A confirmatory action order is intended to confirm a voluntary licensee commitment to specific action and may involve a compliance backfit.

(13) *What about a notice of violation requesting a description of the licensee's corrective action or staff requests for licensees to consider certain additional actions. Are these backfits?*

A notice of violation requesting a description of a licensee's proposed corrective action is not a backfit. The licensee's commitments in the description of corrective action are not backfits.* A request by the staff for the licensee to consider some specific action in response to a notice of violation also is not a backfit. If the staff is not satisfied with the licensee's proposed corrective action, however, and requests that the licensee take additional actions, those additional actions are a backfit, unless they are an applicable staff position.

Discussions during enforcement conferences and responses to the licensee's requests for advice regarding corrective actions are not backfits.

Definitive statements to the licensee directing a specific action to satisfy staff positions are backfits, however, unless the action is consistent with an applicable regulatory staff position.

(14) *Can bulletins and generic letters be applied in all respects to every facility or will there be cases where the plant-specific backfit process is to be used?*

_____

*Generally, adequate corrective actions would be required pursuant to 10 CFR 50, Appendix B, Criterion XVI.

NRC bulletins undergo generic review by the CRGR. Therefore, it is not necessary to apply the plant-specific backfit review process to the actions requested in a bulletin (see the response to Question 3.1(8) for further discussion). If the staff expands the actions requested in a bulletin for a specific licensee, however, such expansion is considered a plant-specific backfit.

(15) *What happens when a review concludes that a licensee's program in a specific area does not satisfy a regulation, license condition, or commitment?*

Where the staff previously accepted the licensee's program as adequate, any staff-specified change in the program would be classified as a backfit.

For example, in the case of a plant with an operating license, once the SER is issued signifying staff acceptance of the programs described in the safety analysis report (SAR), the licensee should be able to conclude that its commitments in the SAR satisfy the NRC requirements for a particular area. If the staff were to subsequently require that the licensee agree to additional action other than that specified in the SAR for the particular area, such action would constitute a backfit. In the case described in the question, it is likely that the compliance exception in 10 CFR 50.109(a)(4) would apply (i.e., it would be a compliance backfit).

A somewhat different situation exists when the licensee has made a commitment to a specific course of action and the staff has not yet responded. If the licensee has moved ahead in the intervening time to implement the actions the licensee proposed in the submittal and the staff has failed to provide a timely response, the staff position may be considered a backfit. Thus, if a licensee has implemented a technical resolution intended to meet an applicable regulatory staff position, and the staff for an extended period simply allows the licensee resolution to stand with tacit acceptance, indicated by nonaction on the part of NRC, a subsequent action to change the licensee's design, construction, or operation is a backfit.

3.4 Backfit Analysis

(1) *When is a backfit analysis needed?*

A backfit analysis is needed when a new staff position or legal requirement goes beyond what is necessary for adequate protection and it is not needed to bring a facility into compliance. In other words, if the proposed action would provide a substantial enhancement to safety, the backfit rule provides a

mechanism for imposition as long as the cost of the new position or requirement can be demonstrated to be justified.

(2) *There is increasing evidence of the Commission's apparent willingness to accept subjective cost/benefit analysis. How can such analyses be consistent with the backfit rule?*

The backfit rule requires an analysis. This analysis will vary depending on the nature of the issue, the extent and type of information available, and the ease to which a complex situation can be analyzed by either quantitative or qualitative factors. In some cases, the Commission makes decisions on the basis of qualitative factors. Some of the factors to be addressed in the backfit analysis are not easily quantifiable. In addition, the rule includes consideration of other "relevant and material" factors, some of which may be qualitative.

Quantitative factors, where known, are used, but need not be the only basis for approving a backfit analysis. The complexity and comprehensiveness of the analysis should be appropriate (limited) to what is necessary to provide an adequate base for making a decision. Section III, A.2, "Scope of the Analysis," of NUREG/BR–0058 states: "The emphasis [in doing the analysis] should be simplicity, flexibility, and common sense, both in terms of the type of information supplied and in the level of detail provided." All backfit analyses are to be approved by the cognizant office director or regional administrator and, if generic requests or requirements are involved, a further review by CRGR is also necessary.

(3) *In issuing Generic Letter 88–01, "NRC Position on IGSCC in BWR Austenitic Stainless Steel Piping," why did the staff not consider plant-specific differences that might affect the conclusion of the generic cost-benefit analysis?*

Generic Letter 88–01 was issued on the basis of ensuring compliance with existing regulations; thereby meeting one of the exceptions in the backfit rule for not performing a backfit analysis. Accordingly, a cost-benefit analysis was not required for this action.

In the case of a cost-justified substantial safety enhancement, the costs are analyzed, a finding that they are justified is made by the cognizant office director and is further evaluated by the CRGR. Plant-specific differences and the associated cost information are used when such information is available. In these cases, the estimated range of costs is used by the committee in its review. The overall impact on the industry is generally known with reasonable accuracy, but the development of a specific and

18

detailed cost estimates for each facility is generally not practical.

3.5 Appeals

(1) *Sometimes a licensee will state that a specific action looks like a backfit, but will choose not to pursue the issue at the time. Is the inspector required to do anything?*

No, the inspector need not take any action. Manual Chapter 0514 provides a mechanism for licensees to file a backfit claim whenever they believe an unidentified backfit is imposed on them. The procedure provides guidance on how to file and to whom the claim should be addressed. There are no provisions for verbal claims and no action would be required of inspectors in the circumstances posed unless the claim is filed in writing. Licensees should be advised to file a claim in writing in accordance with the procedures of Manual Chapter 0514.

(2) *How does the staff handle a situation where in response to enforcement action the licensee claims that a backfit is involved? For example, some licensees are claiming backfit in responses to the 10 CFR 50.49 rule. How should any inspector handle such claims particularly if they involve a generic requirement?*

Backfit claims need to be in writing and each claim should be handled on its individual merits by the office responsible for the requested action. At this time, no specific appeal process has been established for generic issues that have been evaluated and reviewed by the CRGR and/or implemented by the Commission. Thus, in such cases licensees should be advised to implement claims of improper backfit in accordance with the plant-specific procedures contained in Manual Chapter 0514.

(3) *In response to a notice of violation, if the licensees use the word "backfit," is the agency required to respond?*

As with any response to a notice of violation, the staff will review and act, as appropriate, on the licensee's response. However, indicating that an unauthorized backfit may have occurred in a response to a notice of violation does not constitute a proper backfit claim that initiates the appeal process set out in Manual Chapter 0514. While the NRC may consider the information surrounding the licensee's claim, no specific agency action is required unless the licensee files a formal backfit claim in accordance with Manual Chapter 0514.

(4) *Does the utility have to formally say it is filing a claim of backfit under 10 CFR 50.109? With whom does the licensee file the backfit claim?*

Manual Chapter 0514 states that a written claim of backfit should be sent to the appropriate office director or regional administrator with supporting rationale and backup information. There is no need to reference the rule since the manual chapter exists to implement the backfit rule. However, the claim must be in writing and should clearly state its purpose.

(5) *Some individuals believe that the staff is attempting to impose new standards on the industry through enforcement. How can licensees use the backfit process to resolve these issues?*

The benchmark which the staff uses in requiring licensee actions is the assurance of safety. If there is a disagreement between the staff and the licensee on what actions are necessary to ensure an adequate level of safety, it can usually be resolved through discussion. If the staff requires actions beyond applicable regulatory staff positions, a backfit would seem to be involved. The nature and justification for backfitting actions are to be consistent with 10 CFR 50.109 and relevant staff guidelines. For further information, refer to the appeal process in Manual Chapter 0514.

3.6 General Questions

Why is there a backfit standard in senior executive service (SES) contracts? Doesn't that send a message that there should not be backfits?

The backfit standard in SES contracts holds NRC managers responsible for proper implementation of the backfit process. SES contracts also contain a standard on ensuring safety, which is of overriding importance and works with the backfit standard. The purpose for including the backfit standard is to emphasize to the staff that management needs to be aware of and to control staff activities to ensure the agency's adherence to the backfit rule and Manual Chapter 0514. Backfits are expected, but they should be properly identified as backfits and handled in accordance with specified procedures. The purpose of instituting controls is to eliminate unauthorized backfits, and the preparation of the SES contract item is to help ensure appropriate management review and oversight over these controls.

APPENDIX A
THE 1985 BACKFIT RULE

Rules and Regulations

Federal Register

Vol. 50. No. 183

Friday. September 20. 1985

This section of the FEDERAL REGISTER contains regulatory documents having general applicability and legal effect, most of which are keyed to and codified in the Code of Federal Regulations, which is published under 50 titles pursuant to 44 U.S.C. 1510.
The Code of Federal Regulations is sold by the Superintendent of Documents. Prices of new books are listed in the first FEDERAL REGISTER issue of each week.

DEPARTMENT OF AGRICULTURE

Food Safety and Inspection Service

9 CFR Part 381

[Docket No. 83-007F]

New Turkey Inspection System

Correction

In FR Doc. 85-22062, beginning on page 37308 in the issue of Monday, September 13, 1985, making the following correction:

On page 37313, first column, the section number in the section heading, which reads "§ 318.76", should read "§ 381.76".

BILLING CODE 1505-01-M

NUCLEAR REGULATORY COMMISSION

10 CFR Parts 2 and 50

Revision of Backfitting Process for Power Reactors

AGENCY: Nuclear Regulatory Commission.

ACTION: Final rule.

SUMMARY: The Nuclear Regulatory Commission is revising its regulations to establish standards and an agency discipline for future management of backfitting for power reactors. Backfitting is a process which can include both plant-specific changes and generic changes as applied to one or more classes of power reactors. As described in the rule, backfitting is defined as the modification of or addition to systems, structures, components, or design of a facility; or the design approval or manufacturing license for a facility; or the procedures or organization required to design, construct or operate a facility; any of

which may result from a new or amended provision in the Commission rules or the imposition of new or different regulatory staff position interpreting the Commission rules after (i) the date of issuance of the construction permit (CP) for the facility for facilities having construction permits issued after October 21, 1985; or (ii) six months before the date of docketing of the operating license (OL) application for the facility for facilities having construction permits issued before October 21, 1985; or (iii) the date of issuance of the operating license for the facility for facilities having operating licenses; or (iv) the date of issuance of the design approval under Appendix M. N, or O of 10 CFR Part 50.

EFFECTIVE DATE: October 21, 1985.

FOR FURTHER INFORMATION CONTACT: James R. Tourtellotte. Chairman. Regulatory Reform Task Force, U.S. Nuclear Regulatory Commission, Washington, DC 20555. Phone: (202) 634-3300.

SUPPLEMENTARY INFORMATION:

Background

The Commission initiated this rulemaking proceeding for the purpose of establishing requirements for the future management of backfitting for power reactors. Backfitting can include both plant-specific changes and generic changes applicable to one or more classes of power reactors.

Section 50.109 of the Commission's current power reactor regulations provides the following standard for backfitting decisions: Backfitting may be required where the Commission finds "that such action will provide substantial, additional protection which is required for the public health and safety or the common defense and security." On its face, this appears to be a relatively high standard. In practice, however, § 50.109 has rarely been formally invoked, and it is therefore difficult to tell the extent to which this standard has actually been applied to previous backfitting decisions. The Commission has decided that a new, more specific standard and related procedures should be applied by rule to backfitting decisions.

The Commission published an advance notice of proposed rulemaking and policy statement on this subject at 48 FR 44217 (September 28, 1983) and more recently, a notice of proposed

rulemaking at 49 FR 47034 (November 30, 1984). The complete record of this proceeding is available for review in the Commission's Public Document Room at 1717 H Street, NW., Washington. DC.

Public Comments

The comment period officially closed January 29, 1985. A number of comments were received after that time, the last of which was filed on March 12, 1985, by the Advisory Committee on Reactor Safeguards. All comments were considered in formulation of the final rule.

Fifty-seven comments were filed as follows: utilities, 30; vendors, 3; architect engineers and service companies. 5; industry groups and trade associations. 3; consulting engineering firms. 3; various individuals and groups, 10; federal agency, 1 (DOE); states, 1 (Illinois); Advisory Committee on Reactor Safeguards, 1.

As a result of the responses to the advance notice of proposed rulemaking, the Commission posed six numbered questions and other unnumbered questions in the notice of proposed rulemaking. The responses to these questions have assisted the Commission in reaching its determination on the content of the final rule.

Question 1. Should § 50.109 also apply to backfitting imposed through rulemaking? When a modification is imposed by rule or regulation. should the affected licensee be afforded an appeal to the EDO? What is the basis for this position?

The Union of Concerned Scientists (UCS) stated that § 50.109 should not apply to rulemaking. They assert that the Atomic Energy Act and prevailing case law do not permit the consideration of cost in determining minimum safety standards. (See UCS 1983 comments, pages 4-7.) An appeal to the EDO from a requirement imposed by rule cannot be legally permitted, according to UCS, and the Commission may not circumvent the legal requirements of the Administrative Procedure Act. 5 USC Section 553, by permitting appeals outside of the public forum to the Executive Director for Operations.

The Ohio Citizens For Responsible Energy (OCRE) also oppose application of § 50.109 to rulemaking because they say "licensees are afforded enough opportunities in the rulemaking and administrative process to contest the rules." They suggest that a petition for waiver of a rule under 10 CFR 2.758 or an exemption under 10 CFR 50.12 provides sufficient remedies for licensees.

The Nuclear Utility Backfitting and Reform Group (NUBARG) believes that backfitting controls should apply to facility modifications imposed by rulemaking. They state four reasons for their position. First, in terms of public health and safety, they state the practical impacts of backfitting by rulemaking or backfitting on a plant specific basis are the same. Therefore, NRC regulations should require a documented analysis of a backfit regardless of the source of the requirement. Second, there is no apparent justification for excluding backfit modifications imposed by rulemaking. They suggest that the NRC should satisfy itself of the need for and efficacy of any backfit required. Third, if the backfitting rule did not apply to rulemaking, there may be a natural temptation by the staff to avoid the effects of the backfitting rule by imposing requirements through rulemaking. Fourth, there would be no additional burden because much of what the rule would require already takes place during the CRGR review of proposed rules. NUBARG states that it does not advocate the preparation of a plant-specific backfitting analysis for backfits proposed in the context of a rulemaking.

NUBARG also believes that an opportunity for an appeal to the Executive Director for Operations should exist. The licensee, they say, should be given the opportunity to demonstrate that the modification established by rule or regulation should not be required for its facility because that facility is substantially different from the type, design, or vintage of facilities evaluated in the modification analysis and as a result, findings made pursuant to § 50.109 are not applicable. They go on to cite the need for flexibility in the rulemaking process as a basis for their position. The Atomic Industrial Forum (AIF) and other industry commenters appear to be in general agreement with the positions taken by NUBARG.

DOE also states that § 50.109 should apply to rulemaking since rulemaking and orders are "the only two avenues through which a backfit should be imposed by the Commission." They oppose appeal to the EDO, however, and suggest use of a waiver request under 10 CFR 2.758.

Question 2. Should § 50.109 limit backfitting to backfits imposed by rule, regulation or order? If the imposition of

backfits is not limited to rules, regulations or orders, what other mechanisms should be employed?

UCS opposes such a limitation, stating that the effect would "undoubtedly be to eliminate the condition which the Supreme Court found legally necessary to justify two-stage licensing," citing *Power Reactor Development Company v. Union*, 367 U.S. 396, 414 (1961).

OCRE takes a somewhat different position. They state that the Atomic Energy Act "clearly states that the Commission's safety standards are to be imposed by rule or order . . . However, the NRC is in the habit of imposing regulatory requirements through non-enforceable means (e.g., Reg. Guides, SRP). OCRE states that because legally binding requirements are those reached through rulemaking or adjudication and because these processes inherently involve weighing pros and cons of adverse parties, they are reasoned, open deliberated processes subject to judicial review and therefore need no further analytical requirements. OCRE continues, "While it would be preferable if all regulatory requirements resulted from rules or orders, it is a fact of life that the staff imposes regulatory requirements on its own." OCRE appears to not take a position either way on the question but is willing to accept current staff practice as a "fact of life."

NUBARG takes a strong position that § 50.109 should limit backfitting to those modifications imposed by rules, regulations or order. They state that current and past staff practice of requiring licensees to backfit facilities on the basis of non-binding guidance requirements is illegal. Regarding the second part of the question, NUBARG states that there are no means other than rules, regulations or orders by which the Commission may lawfully require a licensee to modify its facility. In short, it is NUBARG's position that Section 50.109 would violate the statute if it permitted imposition of backfits by any means other than rules, regulations or order. The AIF and other industry commenters appear to be in general agreement with NUBARG's position.

DOE states that backfitting should only be imposed by rule, regulation or order and that all analyses, reviews and decisions required by the proposed rule should apply to all methods of backfitting.

Question 3. Should a documented analysis of a proposed backfit come before the backfit is issued or only after an affected licensee lodges an appeal?

USC urges that there be no requirement for a detailed analysis unless the licensee appeals because such

analyses in absence of an appeal would, in their words, "be an utter waste of time and resources."

OCRE suggests that to require analysis of every proposed backfit would create too great a burden on the staff. OCRE appears to reserve the term "backfit" for "non-enforceable regulatory requirements" and therefore, "licensees should feel free to contest a proposed backfit."

NUBARG takes the position that there should be a documented analysis by the NRC whenever its proposes to require licensees to modify their facilities. They state, "a plant modification has the same impact regardless of who initiates it. Therefore, just as the licensee must always develop a sound technical basis in support of a proposed facility modification, so should the staff." Such an analysis is necessary, they argue, so that the NRC can be assured that the backfit it wishes to impose is truly needed to enhance safe reactor operations and that it will have the intended effects.

AIF suggests that such evaluations are needed to determine whether the proposed backfit does increase safety, to what extent, and at what costs. Further, it is needed "to impose discipline into the backfit process." AIF also suggests that licensees should not be placed in a position of having to invoke procedure in order to initiate backfit analysis. To do so, they say, places the licensee in a position of having to jeopardize its relationship with the staff by opposing a change that the staff is requiring.

AIF also suggests that, in addition to the seven factors proposed in the Federal Register notice, the following factors should be considered in making an analysis of a proposed backfit.

1. A precise statement of the specific objectives that the proposed modification is designed to achieve.

2. A general description of the activity that would be required by the licensees or applicants in order to complete the modification.

3. Alternatives to the proposed backfit and how these alternatives (including the recommended alternative) will affect other proposed or imposed facility backfits; and

4. A priority ranking by safety significance of each proposed backfit relative to other proposed or imposed backfits.

5. Whether, after balancing of all appropriate factors (including those in this paragraph) the demonstrations required in § 50.109(a) have been made.

DOE states that the burden of proof for demonstrating that an increase in safety is needed should rest with the staff rather than requiring the licensee to prove that such an increase is unnecessary. Their reasoning is that. "requiring a written basis and analysis of a proposed backfit before it is imposed will increase the likelihood of improved safety and increase confidence that its effects are understood."

Question 4. Should backfitting be defined as "the imposition of new regulatory requirements or the modification of previous requirements" (the cause) or defined as a "modification or addition required by the Commission to the facility or to the structure. systems or components of such facility. the design thereof. or the procedures or organization required to construct or operate such facility" (the effect)? What is the basis for this position?

UCS believes that neither definition is appropriate. citing its 1983 comments. pages 10–30. in support of its position. UCS further suggests that exclusion of rules. regulations and orders from the definition of regulatory requirements raises questions about what is meant.

OCRE states that backfitting should be defined as "the imposition of new requirements; i.e.. the cause. not the effect." Its reasoning is that "Since we interpret backfit to apply *only* to the non-enforceable requirements. licensees are free to use alternative methods to comply. This, again. is a difficult point which should be resolved by bringing the NRC's practices into line with the Atomic Energy Act (AEA); i.e.. all requirements imposed by rule or order."

AIF suggest that the definition should be stated in terms of the effect which they suggest should read:

As used in this section. "backfitting" of a production or utilization facility means a modification or addition required by the Commission to the design approval. manufacturing license. or facility or to the structures. systems or components of such facility. the design thereof. or the procedures or organization required to construct or operate such facility. after . . . (times specified in proposed rule).

They also suggest that backfits should include requirements to perform extensive analytical efforts or tests. regardless of whether modifications or additions to the structures. systems or components of a facility or design result from such analytical efforts or tests. The basis for rejecting regulatory requirements as a part of the definition is directly related to their previous argument that backfits may only be legally imposed by rule. regulations or order. AIF's position is generally supported by other industry comments.

DOE would recommend the following in lieu of either the "cause" or "effect" definition:

1. A "modification." means a change required by the Commission to a site permit: a design approval: a production or utilization facility. or the structures. systems. or components of a facility; to the procedures pursuant to which a facility is to be constructed or operated: or to the organization required to construct or operate such a facility.

2. A "backfit" means "a modification not imposed by the Commission for achieving compliance with a construction permit or operating license. at the time of issuance or as amended. or contained in the requirements incorporated by reference in the permit or license."

The State of Illinois rejects the use of "regulatory requirements" as too ambiguous and suggests the definition be more precise for clarity and scope.

Question 5. The industry's proposed standard for justification of a backfit is "substantial improvement in the overalll safety of the plant considered over its remaining life." Is it appropriate to include the concept of "over its remaining life"? What other standard could be used?

UCS suggest that it is not appropriate to consider the concept of "over its remaining life" for the following reasons:

1. Such analysis can only be based on probabilistic risk assessment and that methodology is not appropriate.

2. The concept creates an incentive for delay and obstructionism and rewarded those who delay the most.

3. "Benefits" are currently expressed by NRC in terms of annual average dose "avoided" and this is inconsistent with the concept of "remaining life".

4. The concept does not account for problems cause by aging and deterioration of equipment which are likely to increase as a plant ages.

5. There is no justification in law or policy for subjecting people around older plants to a greater risk than those who live around newer plants.

OCRE also objects to the use of the standard because of what they perceive to be implication of required use of probabilistic risk assessments.

NUBARG suggests that use of the concept is appropriate as being one factor among many that should be considered when a backfit is required. Industry commenters generally support this position.

Question 6. To what extent may the Commission consider cost. including the economic costs in backfitting decisions under standards and processes proposed in § 50.109?

USC cites its previous 1983 comments in support of its position that costs may not be considered under the Atomic Energy Act and established case law. OCRE also opposes cost consideration as a part of the decision process.

AIF takes the position that cost may be considered and that such costs should include:

1. Costs of evaluation. engineering. construction. material procurement. Allowance for Funds Used During Construction. (AFUDC) and investigations:

2. An attributable portion of replacement power costs during down time for implementation:

3. Operating costs due to changes in specifications. procedures. operator retraining and training manuals. increases in manpower requirements and net generation losses:

4. Impact on preoperational startup. operator training. procedure development and system turnover during plant construction: and

5. Any incremental increase in man rem exposure as a result of installation and subsequent operation of the modification.

As a basis for the position stated. AIF attaches to their comment a legal memorandum entitled. "Consideration of Cost and Benefits in Connection with Backfitting." This memorandum takes the position that the Atomic Energy Act and its legislative history. court decisions. Commission regulations and documents. the Energy Reorganization Act and Executive Order 12291 and the NRC's General Counsel memorandum dated May 4. 1984. all support the conclusion that costs may be considered in connection with backfitting. Industry comment generally supported the AIF position.

DOE also conducted a legal analysis of the cost question. They stated:

The legal conclusion which emerges from the foregoing is that. except for deciding the narrow question of whether a backfit should be required for construction permittees to eliminate or reduce to a threshold level a particular risk in order to meet the "adequate protection" test. the NRC has broad discretion to consider the relationship between benefits and costs in deciding whether to impose a backfit.

The Commission also requested comments on whether reliance upon probabilistic risk assessments is prohibited by the Atomic Energy Act as suggested by UCS. OCRE agreed with the UCS position.

AIF takes the position that UCS mischaracterizes the industry position on the use and value of probabilistic risk assessment (PRA). They point out that

PRAs should support. not supplant. determinative requirements. NUBARG points out that neither the industry nor the proposed backfitting rule mandates the use of PRAs. They point to the fact that the proposed industry rule would require the use of PRAs only "where appropriate and where pertinent data is available." They also suggest that the Atomic Energy Act does not prohibit the use of PRAs.

The Commission requested comments on the correctness of the UCS position that "the Commission exercises its rulemaking authority to establish nuclear reactor safety standards, and licenses may avoid those standards only by obtaining a waiver under 10 CFR 2.758." NUBARG states that UCS misunderstands § 2.758 and the operation of the backfit rule. They further suggest that neither of the proposed backfitting rules can reasonably be read as permitting licensees to avoid requirements applicable to their facilities. Those rules, they state. would simply require the staff to document the basis for its conclusion that a backfit is required.

The Commission requested comments on whether the elements of the proposed backfitting rule are too prescriptive and are truly needed to ensure that the staff considers all factors that are appropriate before it imposes a backfit. NUBARG points out that virtually all of the elements of the analysis have been used by NRC before and are sufficiently broad to be applied in most, if not all, by backfitting situations. The State of Illinois remarked: "The Department [State] believes that the seven factors contained in the proposal provide an appropriate means for balancing all factors in determining whether backfitting should apply." AIF agreed with the seven factors but suggested the addition of five more.

The Commission also expressed a concern over whether preparation of a ··ackfitting analysis should be required as a condition precedent to the issuance of a license amendment. NUBARG stated that "unless requested by a licensee, the staff should not be requested to prepare a backfitting analysis as a condition precedent to issuance of a license amendment if the licensee requests an amendment pursuant to 10 CFR § 50.90." NUBARG points out that application for significant amendments requries a description of the proposed modification and the preparation of a safety analysis report by the licensee. Since the licensee presumably will have subjected the amendment to an internal cost effectiveness review. a backfitting

analysis by the NRC would appear to be neither necessary nor appropriate. AIF was in general agreement with this position and stated further that the option to allow a licensee to request a backfitting analysis should be retained. AIF suggested that there are instances when licensees are under informal but intense regulatory pressure to submit an amendment request. In this circumstance. backfitting analysis should precede the issuance of a license amendment according to AIF. General comments from other members of the industry tend to support the NUBARG and AIF positions.

Comments on the Additional Views of Commissioner Asselstine

Commissioner Asselstine's additional views were generally supported by Ecology/Alert. Federal Conservationist of Westchester County, Inc.. Ohio Citizens for Responsible Energy, and UCS. Industry comment generally opposed Commissioner Asselstine's approach. Similarly, the Department of Energy did not support Commissioner Asselstine's alternative backfit rule, and the State of Illinois had a mixed response.

Although UCS endorses Commissioner Asselstine's position. it suggests two changes. First, it takes exception to Commissioner Asselstine's rule to the extent that it prohibits consideration of monetary costs at the operating license stage only for backfits related to safety matters that were left unresolved at the time of issuance of the construction permit. UCS believes that so long as construction permits are to be granted on the basis of preliminary design concepts. it is not legitimate to consider as backfits. changes required between the construction permit and operating license, or to consider costs at that stage. Second. UCS objects because Commissioner Asselstine's proposal does not provide for formal public participation in backfitting decisions. USC believes that the decisionmaking process should be open and accessible to all persons who might be affected.

OCRE also suggested two changes to Commissioner Asselstine's proposal. First, they would remove review by CRGR because. they say. CRGR does not further the mission of the Commission but serves only to discourage new safety improvements. Second. they. like UCS. would provide an opportunity for public comment for both generic and plant specific backfits.

In its discussion rejecting the proposed use of "regulatory requirements" in the definition of backfitting. the State of Illinois endorses "the more precise definitions of ·

backfitting proposed by Commissioner Asselstine and the industry" and to that extent. could be considered as endorsing Commissioner Asselstine's approach. However. the State of Illinois also stated that they did not agree with Commissioner Asselstine's proposals to specify in the regulations a presumption in favor of the backfit. They believe that seven factors contained in the proposal provide an appropriate means for balancing all factors in determining whether backfitting should apply.

The thrust of the industry position appears to be that many of the terms used by Commissioner Asselstine in his proposed rule are ambiguous and undefined and in other instances, where the standard is well understood. it is simply misconceived. For example. NUBARG points to Commissioner Asselstine's proposal to define backfits in terms of changes to facility design. construction or operation "imposed by the staff to . . . satisfy a regulatory staff position" developed after a specified period. NUBARG complains that "regulatory staff position" is not defined. AIF states that the word "satisfy" in this context cannot be anchored to any applicable statutory standard. not to any prevailing doctrine of administrative jurisprudence. NUBARG questions the ultimate effectiveness of such an alternative rule because. they argue. backfits may not be legally imposed on the basis of such documents.

Industry takes a different tack with regard to the position espoused by Commissioner Asselstine that the basic premise of nuclear regulation should be to "reduce the risk to the public caused by these facilities to a level that is as low as reasonably achievable." NUBARG suggests that this approach reverses the presumption of regularity associated with past NRC licensing decisions. Those who have already been granted licenses and thus have been deemed "safe enough" by the NRC could. according to NUBARG. find themselves having to justify routinely why their licenses should not be modified. This, NUBARG states. raises serious legal questions of fundamental fairness and due processs, and appears to be at odds with the Administrative Procedure Act. NUBARG also complains that the standard suggested by Commissioner Asselstine is potentially open-ended.

AIF further suggested that the Atomic Energy Act requires "reasonable assurance of the public health and safety" and reasonable assurance is not equated with "as low as reasonably achievable." AIF further states that this standard is at odds with section 103(b) of the Atomic Energy Act, which provides, in part, that "The Commission shall issue such licenses . . . to persons applying therefor . . . (to) who are equipped to observe and who agree to observe such safety standards to protect health and to minimize danger to life or property as the Commission may, by rule, establish: . . ." AIF suggests that this language has been interpreted by the Commission in its regulations to require "reasonable assurance" that licensed activities of the Commission can be conducted without endangering the health and safety of the public, citing, for example, 10 CFR 50.57(a)(3). They also cite *Citizens for Safe Power, Inc.* v. *Nuclear Regulatory Commission*, 524 Fed. Second 1291, 1297 (D.C. Circuit 1975) for the proposition that "absolute or perfect assurances are not required by AEA and neither present technology or public policy admit of such a standard."

The Department of Energy also does not support Commissioner Asselstine's alternative proposed backfit rule. This proposed rule, DOE states, "detracts from the basic purpose for instituting a new backfit rule and, if adopted, would perpetuate the significant deficiencies of backfitting practices of the past." DOE further suggests that Commissioner Asselstine's definition of backfitting is too narrow; that the "as low as is reasonably achievable" standard is inappropriate, and would probably be inconsistent with safety goals should those be established; that the limitations on the use of quantitative cost benefit balancing would be "overly restrictive" and would be "a regressive step for modern analysis techniques"; that the decision criteria are not identified in Commissioner Asselstine's rule; and that the implementation procedures have several deficiencies.

Commission Position

The Commission is appreciative of the time and effort expended by those who submitted comments. Backfitting is a matter of considerable importance and the views expressed in the comments have been very helpful to the Commission in its deliberation. To some extent, the final rule will be modified from the proposed rule to reflect the views expressed.

Since there is no practical difference between a backfit that is imposed pursuant to a rule or a staff position interpreting a rule, the Commission will alter the final rule to require a documented analysis of required backfits regardless of the source. A plant-specific backfit analysis will not be required in rulemaking and the

factors specified in the rule will be reviewed only on a generic basis for rulemaking purposes. Because there must be safety reasons for the agency to impose any changes to a regulatory requirement or a staff position, applicable to the licensee, because the safe consequences are unknown until analyzed, and because the Commission should fully understand the effects of a proposed backfit before its imposition, it is of little consequence how a backfit is imposed. Safety and sound management require that analysis precede imposition of a new or modified regulatory requirement or staff position. It follows that those backfits imposed by rulemaking should undergo the same scrutiny as proposed by other means. It also follows that changes in regulatory requirements or staff positions for procedures and organization should also be analyzed before implementation to determine, *inter alia*, the safety significance of any such proposed change. The final rule reflects this position.

Many of the most important changes in plant design, construction, operation, organization, and training have been put in place at a level of detail that is expressed in staff guidance documents which interpret the intent of broad, generally worked regulations. The NRC has determined that the correct focus for backfit regulation is the establishment of effective management controls on existing staff processes for the interpretation of regulations that are known to result in valuable upgrades in industry safety performance. Thus, the Commission opts to adopt a management process not only for the promulgation of regulations as backfit instruments, but also for the lower tier staff review and inspection processes known to result in reactor plant changes.

The Commission agrees with those who suggest that the Staff should not be required to prepare a backfitting analysis as a condition precedent to issuance of a license amendment if the licensee requested the amendment pursuant to 10 CFR 50.90. If a licensee believes that the amendment process is being used by the staff to impose a backfit, the licensee may invoke the rule under § 50.109. It is unnecessary to amend the rule in this regard since mention of the point here provides adequate direction to the Staff and licensees.

Considerable attention was given to the question of whether backfitting should be defined in terms of its cause or its effects. After due consideration, the Commission believes that the definition for backfitting should take

into account both the cause and the effects. Therefore, the definition is modified accordingly.

Question 5 concerned the industry's proposed standard for justification of a backfit and the suggestion that the "substantial improvement in overall safety of the plant considered over its remaining life" should be incorporated into the rule. In our view, the concept of "over its remaining life" is already incorporated in the rule under § 50.109(d)(8). There is no need to place that concept in the rule at another place.

The additional factors suggested by the industry for inclusion under § 50.109(c) generally appear to be reasonable and not unduly burdensome. Therefore, the thrust of the additional factors will be included as appropriate in the final rule.

As the accountable manager for backfitting, the Commission has directed the EDO to establish backfit procedures and to ensure appropriate rights of appeal. The Commission believes it is unnecessary to include in the rule a section establishing appeal rights to the Executive Director for Operations.

Consideration of Costs in Backfit Decisions

In the current rulemaking, comments were filed by UCS and AIF stating strongly contrasting legal views concerning the Commission's authority to consider the costs of new safety requirements which the Commission would impose if costs were not a factor in the decision. (See Question 6, *supra*.) In view of the importance of the cost issue and the strongly divergent views stated in the comments, it is important to set forth the Commission's legal and policy views on this matter.

The costs associated with proposed new safety requirements may be considered by the Commission provided that the Atomic Energy Act finding "no undue risk" to the public health and safety can be made. There may be any number of ways by which the Commission can arrive at such a conclusion. Each approach could have different costs associated with it and it cannot be seriously argued that in such circumstances the Commission is statutorily prevented from choosing the most cost effective means of protecting public health and safety.

Similarly, it may be presumed that the current body of NRC safety regulations provides adequate protection. Where new information indicates that improvements are needed to ensure there is "no undue risk" on either a plant-specific or generic basis which the Commission believes to be the minimum necessary, such requirements must be imposed. However, where there are alternatives for achieving the improvements which have different associated costs, such costs may be considered.

Cost considerations have been a part of the Commission's regulatory approach in many other instances. For example, the ALARA principle requires Commission licensees to meet an absolute set of radiation exposure standards but also requires further reductions in exposure where the cost of the exposure avoided outweighs the cost of implementing controls to avoid the exposure. Commenters who addressed the proposed backfit rule and opposed the use of costs did not address this point. It would appear that the only situation where the consideration of costs may be seriously challenged is where a new requirement is necessary to provide an absolutely minimum level of protection to the public health and safety and no alternative means of achieving such protection are apparent.

In general, the consideration of costs associated with incremental safety improvements is within the NRC's statutory mandate. However, the cost of new safety requirements will not be considered where such requirements are necessary to ensure there is no undue risk to the public health and safety and no alternatives are available.

After reviewing all of the comments and positions stated, the Commission believes that there is sufficient authority in the statutes, case law, and Commission practice to justify making cost considerations in backfitting decisions. Since consideration of costs was a part of the proposed rule, the rule will remain unchanged in this regard. The Commission also rejects as without merit the suggestion that probabilistic risk assessments are precluded by law.

Description of Final Rule

The proposed amendment of § 50.54(f) ensures that except for information sought to verify licensee compliance with the current licensing basis for that facility, the reason or reasons for each information request must be prepared prior to its issuance to determine whether the request is for information already in the possession of the applicant or licensee or instead will require the institution of studies, procedures, or other extensive effort to generate the necessary data to respond. If extensive effort is reasonably anticipated, the request will be evaluated to determine whether the burden imposed by the information request is justified in view of the

potential safety significance of the issue to be addressed.

It should be noted that § 50.54(f) does not by its terms apply to the review of applications for licenses or amendments. Consequently, if the staff seeks information of a type routinely sought as a part of the standard procedures applicable to the review of applications, no analysis will be necessary. If the request is not part of routine licensing review and falls within the purview of § 50.109, however, a full analysis is most likely indicated. Requests for information to determine compliance with existing facility requirements or for fact-finding reviews, inspections and investigations of accidents or incidents, however, usually are not made pursuant to § 50.54(f) nor are such requests normally considered within the scope of the backfit rule. Amendment of this section also provides for management control and accountability for backfits by requiring that staff evaluations be reviewed by the Executive Director for Operations or his designees prior to the issuance of the request.

The amendment of § 50.54(f) should be read as indicating a strong concern on the part of the Commission that extensive information requests be carefully scrutinized by staff management prior to initiating such requests. The Commission recognizes that there may be instances where it is not clear whether a backfit will follow an information request. Those cases should be resolved in favor of analysis. In short, staff management should develop an internal review process to ensure that there is a rational basis for all information requests, even where it is not clear that a backfit will result.

Section 50.109(a) sets out the definition of backfitting, the analysis requirement, the standard to be used in determining whether a backfit should be imposed and the exceptions to the rule. The definition focuses on modifications to systems, structures, components, designs, procedures or organization which may be caused by new or modified Commission rules or orders or staff interpretations of Commission rules or orders.[1] Thus, this definition includes both cause and effect of backfitting. It may also be noted that "cause" includes not only Commission rules and orders, but staff interpretations of those rules and orders. This is not to say that staff interpretations of rules are viewed by the Commission as being legal requirements. Clearly, they are not.

---

[1] The term "regulations" is not in the text because that term is synonymous with "rule."

Nevertheless, staff interpretations of broadly stated rules are often necessary to give a rule effect and in some instances may be a causal factor in initiating a backfit.

Section 50.109(a)(2) requires a systematic and documented analysis as a condition precedent to the imposition of a backfit. This will ensure that the safety significance of any modification and its relation to other relevant factors is well understood before changes are required.

The standard against which proposed backfits would be measured is stated in § 50.109(a)(3) as "substantial increase in the overall protection of the public health and safety or the common defense and security." Substantial means "important or significant in a large amount, extent, or degree." Under such a standard, the Commission would not ordinarily expect that safety improvements would be required as backfits which result in an insignificant or small benefit to public health and safety or the common defense and security, regardless of the implementation costs. On the other hand, the standard is not intended to be interpreted in a manner that would result in disapprovals of worthwhile safety or security improvements having costs that are justified in view of the increased protection that would be provided.

The phrase "overall protection of the public health and safety or the common defense and security" in the proposed backfit standard also deserves some discussion. The principal purpose of requiring consideration of the overall protection that would be provided by a proposed backfit is to ensure that both its negative and positive effects are taken into account in deciding whether the backfit is justified. A backfit for a part of a plant should be evaluated in light of the net increase in overall protection that the entire plant would provide as a result of the backfit, taking into account the effects it would have on other aspects of the plant. Thus, the net benefit of a backfit to the protection provided by the plant as a whole is the overriding consideration, not just the benefit to the part of the plant being backfitted.

However, the Commission does not intend use of the phrase "overall protection" in the backfit standard to signal a departure from its traditional reliance on defense in depth and diversity for protection of public health and safety. Therefore, safety improvements in one line of defense against undue risk should not be disapproved or approved based solely on the presence or absence of another line of defense to cope with the failure of the first. For example. safety improvements in the integrity of the reactor coolant system should not be dismissed merely because an emergency core cooling system has been provided to protect public health and safety with high confidence in the event that the integrity of the reactor coolant system is lost. On the other hand. such a suggested improvement may be precluded because it does not meet the substantial test. or does not increase overall protection provided by the plant due to. for example. the negative impacts on other aspects of the plant. The proposed requirement that the costs of backfits be considered and justified in view of the increased protection to public health and safety or security is based on the Commission's view that it should. in these circumstances. consider the direct and indirect costs of implementation in making safety decisions under the Atomic Energy Act.

The consideration and weighing of costs contemplated by the rule applies to backfits that are intended to result in incremental safety improvements for a plant that already provides an acceptable level of protection. In this area the Commission believes that direct and indirect implementation costs are especially relevant. Without cost as a competing consideration in these circumstances. the regulatory process takes on the characteristics of a quest for a risk-free plant. an unattainable objective as recognized by Congress in establishing the standard of no undue risk in the Atomic Energy Act.

Section 50.109(a)(4) creates exceptions for modifications necessary to bring a facility into compliance or to ensure through immediately effective regulatory action that a licensee meets a standard of no undue risk to public health and safety. In cases involving the compliance exception. backfit analysis is not required and the standard does not apply. The compliance exception is intended to address situations in which the licensee has failed to meet known and established standards of the Commission because of omission or mistake of fact. It should be noted that new or modified interpretations of what constitutes compliance would not fall within the exception and would require a backfit analysis and application of the standard.

The exception for immediately effective regulatory actions that are necessary to ensure that a licensee meets the standard of no undue risk to the public health and safety permits the Commission to act in

emergency situations to ensure that operation of the backfit rule will not preclude the Commission from ensuring that minimum standards are met to protect public health and safety. The exception anticipates the existence of significant new information or the occurrence of an event which clearly demonstrates that the standard of no undue risk to the public health and safety cannot be maintained without the designated modification. Moreover, the presumption of safety which ordinarily accompanies the issuance of any license must be overcome in order for the exception to be used. As with the compliance exception. there is no intent on the part of the Commission to include within the scope of the exception new or modified interpretations of what constitutes no undue risk to the public health and safety. In such a case. the rule applies. The rule also provides that a backfit imposed by immediately effective regulatory action shall not relieve the Commission of performing an analysis after the fact to document the safety significance and appropriateness of the action taken.

For those modifications which are to ensure that the facility poses no undue risk to the public health and safety and which are not deemed to require immediately effective regulatory action, analyses are required; these analyses, however, should not involve cost considerations except only insofar as cost contributes to selecting the solution among various acceptable alternatives to ensuring no undue risk to public health and safety.

To ensure that the discipline is maintained in the process and that the exceptions do not become the rule. the Commission directs the staff to document each exception. Documentation shall include a precise statement of the specific objectives of and reasons for the modification and the basis for the exception. It may also serve useful regulatory purposes to include such matters as a general description of the activity that would be required by the licensees or applicants in order to complete the modification and the identification by type. design and vintage of the design approvals. manufacturing licenses for production or utilization facilities to which the modification would apply.

Section 50.109(b) "grandfathers" backfits imposed prior to the effective date of this rule.

Section 50.109(c) sets out nine factors to be used by the staff in its backfit analysis. Finally, § 50.109(d) explicitly recognizes the responsibility of the Executive Director for Operations to

manage the Commission's backfitting program in general and requires approval of backfit analyses by the Executive Director for Operations or his or her designee.

As a matter of information. it may be noted that the nine factors in § 50.109(c) have precedent in existing NRC practices as seen in the Regulatory Analysis Guidelines of the U.S. Nuclear Regulatory Commission. NUREG/BR-0058. the approved CRGR Charter and the Commission's approved plan for the management of plant-specific backfitting. SECY-83-321.[9]

The nine factors to be used by the Staff for a systematic and documented analysis are listed under § 50.109(c) and read as follows: "(1) Statement of the specific objectives that the proposed backfit is designed to achieve: (2) general description of the activity that would be required by the licensee or applicant in order to complete the backfit: (3) potential change in risk to the public from the accidental offsite release of radioactive materials: (4) potential impact on radiological exposure of facility employees: (5) installation and continuing costs associated with backfit. including the cost of facility down time for the cost of construction delay: (6) the potential safety impact of changes in plant or operational complexity including the effect on other proposed and existing regulatory requirements: (7) the estimated resource burden on the NRC associated with the proposed backfit. and the availability of such resources: (8) the potential impact of differences in facility type, design or age on the relevancy and practicality of the proposed backfit: (9) whether the proposed backfit is interim or final and. if interim. the justification for imposing the proposed backfit on an interim basis. These nine factors are to be used as balancing mechanisms in the decisionmaking process for backfitting.

During internal review of the rule. a question was raised as to whether licensing action should be withheld during backfit review. The answer is that the rule never contemplated such a withholding. To the contrary, until a backfit analysis is complete. licensing action should continue along a course

---

[9] The Regulatory Analysis Guidelines of the U.S. Nuclear Regulatory Commission. NUREG/BR-0058. is available for inspection or copying for a fee in the NRC Public Document Room. 1717 H Street. NW.. Washington. D.C. This report may be purchased from the U.S. Government Printing Office (GPO) by calling 202-275-2060 or by writing the GPO. P.O. Box 37082. Washington. DC 20013-7082. It may also be purchased from the National Technical Information Service. U.S. Department of Commerce. 5285 Port Royal Road. Springfield. VA 22161.

consistent with normal practice. For clarification of the point, § 50.109(d) was added to the final rule.

Section 50.109(e) emphasizes and codifies the Commission's intent that backfit management is of paramount importance to responsible regulatory practice. Accordingly, the Executive Director for Operations is responsible for implementation of the backfit rule.

It may be noted that the resolution of any backfit case can be by Commission rule or order, or by written commitment of a licensee. Recognition of this point completes the design of the backfit management process and establishes that licensee compliance with approved backfits may be accomplished by voluntary commitment, but that the legal instrument of a rule or order can and will be used if necessary.

The proposal to amend 10 CFR Part 50. Appendix O is necessary to conform Appendix O to the final rule. The amendment provides that information requests to the approval holder regarding an approved design shall be evaluated prior to issuance to ensure that the burden to be imposed on respondents is justified in view of the potential safety significance of the issue to be addressed in the requested information. Each such evaluation performed by the NRC staff shall be in accordance with 10 CFR 50.54(f) and shall be approved by the Executive Director for Operations or his designees prior to issuance of the request.

Section 2.204 is amended to ensure that any order for modification of a license involving a backfit is subject to the provisions of the new § 50.109.

Commissioner Asselstine and Commissioner Bernthal disapprove this final rule. The separate views and comments of Commissioners follow.

Separate Statement of Chairman Palladino

During my tenure as Chairman, I have sought a new system of backfit controls for NRC that would ensure that a backfit is analyzed and that an explicit judgment of its safety and cost consequences is made. This new rule does just that.

Although a previous version of an NRC backfit rule has been on the books, it has rarely been followed. In addition, documentation in too many cases has been non-existent or inadequate to identify and justify the safety and cost consequences of past NRC-imposed backfits.

The steps to this new backfit rule have been deliberate and, I believe, thorough. In 1983 we issued a policy statement announcing interim backfit controls and our intent to conduct

rulemaking to establish a new backfit rule. 48 FR 44173 (1983). At the same time, we published an advanced notice of rulemaking soliciting public comments on various proposals for the long-term management of the backfitting process, both plant-specific and generic. 48 FR 44217 (1983). In November 1984, we published a notice of proposed rulemaking, seeking public comment on an NRC proposal and on a number of specific questions designed to elicit public views on significant issues and alternatives. The Commission held a number of public meetings on backfitting, and received the advice of the Regulatory Reform Task Force and senior agency officials.

The rule that emerged is a good one. Contrary to the claim of Commissioner Asselstine, the rule is not designed to stymie regulation. What the rule requires is an analysis and an explicit judgment that a proposed safety requirement is justified. A Commission concerned about the protection of public health and safety—which this Commission is, Commissioner Asselstine's comments notwithstanding—will find ample freedom to make sound safety decisions based on analysis. Further, the rule provides that its requirements do not apply if a proposed safety measure is needed to assure compliance with NRC requirements or protection against imminent public risk.

It seems somewhat late in the day for a Commissioner now to argue for the first time that this backfit rulemaking, which we initiated almost two years ago, is unnecessary. To my knowledge, when we started, all Commissioners agreed that our experience under the existing NRC backfit rule had not been satisfactory and that backfit controls of some sort were needed. The decision to incorporate controls into a rule will mean that the Commission can be held accountable in the future for how it implements those controls. Further, the decision to adopt a rule also means that modifications of the controls will involve public participation.

Similarly unfounded is Commissioner Asselstine's criticism of the backfit standard in this rule. The Commission gave considerable attention to the standard during the rulemaking and adopted the following explanation of "substantial increase" in protection of health and safety:

Substantial means "important or significant in a large amount, extent, or degree." Under such a standard, the Commission would not ordinarily expect that safety improvements would be required as backfits which result in an insignificant or small benefit to public health and safety or the common defense and

security, regardless of the implementation costs. On the other hand, the standard is not intended to be interpreted in a manner that would result in disapprovals of worthwhile safety or security improvements having costs that are justified in view of the increased protection that would be provided.

I do not believe that this standard can reasonably be criticized as "intended to block new safety requirements."

Commissioner Asselstine simply ignores the words of the rule when he contends that it requires a backfit analysis that is skewed against new safety requirements. Section 50.109(c) provides that "any . . . information relevant and material to the proposed backfit" may be considered in the analysis and, thus, taken into account in the safety decision. This language provides, in my judgement, ample room for Commission reliance on, among other things, the expertise of its staff to supplement other analytical tools in order to provide an adequate basis for a particular backfit decision.

In response to Commissioner Bernthal's statement, I am disappointed that we could not agree on how to apply backfit controls to future rulemakings. I believe that our differences are really very small.

Modifications to plants or plant procedures can result from new or modified NRC requirements adopted by rulemaking. Therefore, future rulemakings should be covered in principle by backfit controls. The alternative would be a system where plant specific backfits are analyzed and documented but generic rule backfits are not. Such an alternative would leave a significant area of backfitting formally uncontrolled without apparent reason.

Moreover, I believe that the Commission would be creating questions without apparent answers if it chose to control plant specific backfits by a backfit rule and generic rule backfits by internal agency management and guidance. By subordinating generic rule backfits to internal agency management for the stated reason of preserving "flexibility," the Commission could be seen as sending the message that it does not wish to be held accountable for the application of backfitting controls to future rulemakings. This outcome could well serve to undermine the agency's efforts to manage backfitting.

Application of the backfit rule will not result in the Commission making unsound safety or backfit decisions in future rulemakings. The main thrust of the backfit rule is to apply analysis, including analysis of costs, before a backfit is imposed. The rule only requires analysis of "such . . . factors as may be appropriate . . ." The decisional standard is only that the costs of the backfit be "justified in view of the protection to public health and safety afforded by the backfit." Further. as noted earlier. the rule provides that its requirements do not apply if a proposed safety measure is needed to assure compliance with NRC requirements or protection against imminent public risk. Thus. the backfit rule provides sufficient flexibility for the Commission in future rulemakings.

Finally. the Commission can suspend the backfit rule in a future rulemaking if there is good reason. Thus. while it is our intent that this rule apply to backfitting that arises from future rulemaking. the final judgment on this issue will rest with the Commission. If it believes that there is good cause. the Commission could state. in the notice of proposed rulemaking for a future rule. that it was proposing not to apply some or all of the provisions of this backfit rule and request public comment on the underlying reasons. If. after considering public comments. the Commission finds good cause. it can so state in the notice of final rulemaking.

I concur in the views of Commissioners Roberts and Zech.

Comments of Commissioner Roberts and Commissioner Zech

Safety is paramount to the execution of our mission. We believe that the backfit rule is entirely compatible with and supportive of this principle. Unmanaged. uncoordinated and inadequately analyzed backfits. on the other hand. are not. There is nothing in the backfit rule which would stand in the way of a Commission action which is needed to protect the public health and safety or in the way of the adoption of policy changes which a majority of the Commission believes are warranted in the circumstances.

We believe that in the execution of our mission to provide reasonable assurance that the public health and safety are protected. we must have in place criteria and a system for the timely application of justified changes in regulatory requirements. The so-called backfitting regulation which has been in place for many years has not completely satisfied this need. Although it established a broad standard. it did not also provide for a system to assure that backfitting decisionmaking to apply the standard is done in a disciplined. systematic manner. The regulatory system needs a backfitting rule which is complementary to our overall regulatory mission and which is practical to implement. Our chief reason for voting

to adopt this backfit rule is that it provides for a disciplined and formalized review of regulatory requirements to assure that there is a rational basis for modifications to a nuclear power plant. The rule. along with its explanatory statement. also provides guidance and direction to the staff regarding backfit management and control. This disciplined approach to backfitting will. in our judgment. improve the overall effectiveness and certainty in the regulatory process. thus enhancing our regulatory mission.

Discipline and management of backfitting do not mean that safety actions which are justified will be obstructed. or that the Commission will not continue to have the discretion to adopt policies and rules which it believes will serve to enhance the protection of the public health and safety. Instead. they mean that attention and priorities will be focused on areas where action is justified to carry out our regulatory responsibilities. Inadequately managed and controlled backfitting. on the other hand. provide no assurance that modifications. individually and collectively. are in the best overall interest of protecting the public health and safety.

We have carefully considered the views of our dissenting colleagues. and although we respect them. we see the matter quite differently. As we have noted. we believe that this backfitting rule serves a vital regulatory need. We see no reason why the important subject of rulemaking which may involve generic backfitting should be excluded from coverage. It is true. as one of our colleagues points out. that rulemaking is subject to the procedures in the Administrative Procedure Act. It is also correct that we have in place a Committee to Review Generic Requirements (CRGR) and have informal practices which. in their totality may. in an individual case. serve the purpose of the rule if everything falls into place properly in a rulemaking proceeding. The chance of this happening. however. is not to us an acceptable substitute for the system which is being put in place as a matter of overall Commission policy in the backfitting rule. And even assuming that. under its present charter and under its incumbent chairman. the CRGR did cover all of the elements of the backfitting rule. this is not the equivalent of published Commission policy which states the applicable criteria and procedures. The policy and system which are set forth in the backfit rule provide a much sounder foundation for Commission control over future

rulemaking requirements than relying exclusively on existing procedures.

We believe that the procedural requirements of the Administrative Procedure Act and the functions of the CRGR are compatible with the objectives of the backfit rule. However. we also believe that application of the rule to rulemaking is necessary to assure consistent application of the backfitting policy under a prescribed uniform system. regardless of whether a proposed change is generic or plant specific.

The APA does not provide such a system. but does. of course. provide for the important procedural requirement of assuring that there is adequate public notice and opportunity to comment. This does not provide the assurance needed that the substantive matters covered in the backfit rule will be considered either by the staff or the Commission in the Commission's rulemaking decision. There is no requirement for anyone to participate in a rulemaking proceeding. and even for those who choose to participate. there is no provision to assure that there will be systematic and comprehensive coverage of backfitting issues. Indeed. it appears unusual for either the staff's recommendation to the Commission or for the final rulemaking decision to address in any detail. if at all. the application of new requirements to existing licenses and applications. A rulemaking proceeding which meets all of the procedural requirements of the APA would not necessarily assure that the subject matter in the backfit rule is indeed covered. This is not surprising because the objectives of the APA and the backfit rule are. as noted. fundamentally different. Furthermore. we are not able to distinguish. for purposes of the objectives of this backfitting rule. between backfitting modifications which are imposed by individuals in specific plant-by-plant situations and modifications which are imposed by a change in a regulation in a rulemaking proceeding. In each instance a disciplined system should be followed to assure that the backfit is fully understood and justified in terms of its safety relationship and its related costs.

If the backfitting policy and system are sound and are needed. as we believe they are. the straightforward way to go about dealing with the backfit problem is to publish a Commission policy. This is what the backfitting rule does.

We do not share the concern of our dissenting colleague regarding the litigative risks because rulemaking is covered. We are informed that that risk should be minimal. But regardless of whether it is or it isn't. if the rule is

Appendix A   9   NUREG-1409

needed. if it makes sense. and if it is a responsible regulatory action on our part. we should be prepared to defend our decisions in its application.

In summary. we support fully the backfit rule. including its coverage of rulemaking. The rule will bring discipline and accountability to the imposition of plant modifications by the Commission by establishing criteria and a system for rational decisionmaking. Without any question. we believe this approach will enhance the quality of our mission to assure that the public health and safety are protected in our licensing and regulatory requirements decisions.

Separate Views of Commissioner Asselstine

In adopting this backfitting rule. the Commission continues its inexorable march down the path toward non-regulation of the nuclear industry. In two previous decisions the Commission found acceptable the present level of risk of a severe accident at the most highly populated site for an operating nuclear plant in this country. See. *Consolidated-Edison Company of New York* (Indian Point. Unit No. 2). CLI 85–6. 21 NRC 1043 (1985). The Commission's decision was made without an adequate explanation or rationale: it was made without an adequate analysis of the issues: and it was made by ignoring the enormous uncertainties in our methods for estimating risks. The Commission then extended that decision to all nuclear plants through its Severe Accident Policy Statement. See. "Policy Statement on Severe Reactor Accidents Regarding Future Designs and Existing Plants." 50 FR 32138 (August 8. 1985). This backfitting rule is another part of the Commission's withdrawal from active regulation of the industry.

The Commission's rule in effect says that nuclear reactor risks are so acceptable and so well understood that the burden of proof for lowering the risk to the public must be placed on the proponent of improved safety even if that proponent is the Commission itself. This optimistic view of the risks posed by nuclear power plants is unjustified. The Commission's adoption of this rule is truly an unprecedented step in the annals of regulation. I can think of no other instance in which a regulatory agency has been so eager to stymie its own ability to carry out its responsibilities. Indeed. the adoption of this rule is the most compelling evidence to date of the Commission majority's open hostility to the regulatory mission of this agency.

I do not believe that there is a need for a formal Commission regulation restricting the Commission's ability to

require safety improvements for nuclear power reactors. My opinion is shared by our reactor safety technical staff. including the Commission's chief safety officer. The Commission's purported reason for promulgating this rule is to add "discipline" to the backfitting process. The Commission can add discipline to the backfitting process without at the same time unnecessarily limiting its own discretion to impose new safety requirements by fettering itself with this rule. Further. by adding layer upon layer of procedures to the backfitting process the Commission has created a lawyer's paradise in which litigation over procedural irregularities may hamper the Commission's ability to impose needed safety requirements.

Even if I felt that a backfit rule were appropriate. I would not support a rule as poorly thought out as is this one. This rule sets a threshold standard for improvements to safety which is much too high given our present understanding of risk and the uncertainties associated with our methods of estimating risk. Further. the factors to be considered in determining whether a backfit should be imposed are skewed against imposing new requirements. In addition. the Commission's determination of risk in the cost-benefit balance is to be based on unreliable risk analyses.

The consequence of this rule is to limit the NRC staff's and even the Commission's ability to identify and correct safety weaknesses at the nuclear power plants in operation and under construction in this country. As a result. these weaknesses are likely to persist until they cause serious operating events or accidents which pose a direct threat to the health and safety of the public. This rule. then. further ensures a continuation of the piecemeal. reactive approach to safety which has been responsible for many of the failures of the past. By this step. the Commission is moving in the wrong direction—a direction that will likely result in further serious operating events. more accidents. and a lower level of safety than that achieved in many more forward-thinking countries in the world. I discuss each of my reasons for opposing this rule in more detail below.

The Nature of the Backfitting Problem

When asked to describe the backfitting problem. most of our licensees point to the substantial number of hardware modifications. procedural changes and human factors improvements which have been required by the NRC in recent years. The bulk of these new requirements can be traced to three sources: the Commission's fire

protection rule: its rule requiring the environmental qualification of electrical equipment: and the Commission's response to the Three Mile Island accident. It is worth noting that each of these broad safety initiatives was adopted by the Commission itself in response to the identification of significant areas of safety vulnerability within the industry.

Typically. the industry does not challenge the need for improvements in fire protection or the need to assure that safety-related electrical equipment will be able to function under serious accident conditions. Nor does the industry deny the need to address the numerous safety weaknesses brought to light by the Three Mile Island accident. Rather. the industry largely focuses its criticisms on the process used to translate those broad areas of needed improvement into specific modification to plant hardware. procedures and operations.

The industry raises five specific complaints. First. our licensees argue that new requirements often fail to define clearly what is expected of the industry. Second. they contend that the implementation of these requirements—the process by which more general directives are translated into specific modifications—is not well managed. In support of this argument. the industry points to some past failures in documenting proposed modifications. in ensuring consistency in making plant-specific implementation decisions. in providing effective management oversight of plant-specific decisions. and in providing a fair opportunity to appeal objectionable staff-proposed modifications. Third. our licensees assert that specific plant modifications are proposed by staff members who have a single narrow area of interest. and little consideration is given to the overall safety impact of the proposed change. Fourth. the industry argues that the staff's implementation process all too often fails to provide a final decision from the staff on the adequacy of the licensee's efforts to comply with a requirement until *after* the licensee has made the modification to its plant. This process. they contend. exposes the industry to second-guessing by the NRC staff and sometimes leads to making repeated modifications to address the same problem. Finally. the licensees argue that the Commission sometimes adopts arbitrary and unrealistic deadlines for the implementation of new requirements. More than anything else. these complaints focus on the management of the backfitting process.

In the literally thousands of backfitting decisions made by the NRC over the past several years I am sure that some examples can be found to support each of these complaints. However. I believe that each of the industry's valid concerns is addressed by the administrative backfitting management improvements already adopted by the Commission.

Why A Rule?

The Commission claims that this rule is necessary to provide a "management process" for the adoption of Commission regulations and staff interpretations of those regulations. Implicit in the Commission's explanation of this rule is a feeling on the part of the Commission that it is necessary to add discipline to the backfitting process. The Commission never clearly explains why if added discipline is necessary they must embed that discipline or management process in a formal agency rule. The Commission also refuses to explain the connection between discipline and the necessity for a substantial threshold that must be surmounted before safety can be improved. Further. the Commission never adequately explains why it cannot accomplish improved management of the backfitting process without at the same time limiting its own discretion to require safety improvements by requiring that Commission rules comply not only with the Administrative Procedure Act. but also with this rule.

By choosing to adopt this rule the Commission admits failure. The Commission admits that it has been incapable of solving administratively whatever problems it sees with the management of the backfitting process. By adopting this rule. the Commission says to the world that it so mistrusts its own ability to act in a sensible manner and it so mistrusts its ability to control the NRC staff that it must have a formal rule which limits the Commission's discretion and which can be used as a bludgeon to control the staff.

The irony of this all is that the administrative actions taken by the Commission to formalize the backfitting process have already been successful in adding discipline to the process and in addressing the valid concerns of the industry. A senior official of one of the utilities involved most actively in the backfit debate recently told me that when his company first expressed its concerns about backfitting all it wanted was to get NRC management to pay attention to backfitting problems. The company simply wanted a brief-written statement for each proposed backfit describing the proposed change and the NRC staff's reasons for requiring the

change. together with the right to appeal to upper NRC management the decisions being made at the lower levels of the NRC staff. Their objective. this official said. is already being achieved by the Commission's internal management directions to the NRC staff.

Whatever "backfitting problem" exists is really a management problem. The Commission's statement of considerations acknowledges that. And. the problem is being corrected independently of this rule. Why then must this rule be promulgated without delay? Apparently it is because the movement to put in place a backfit rule is much like an avalanch—once it starts rolling it cannot be slowed down or changed in course.

Lawyer's Paradise

By embedding the Commission's backfit management process in the form of a rule. the Commission has chosen to formalize a process which ought to be a purely internal management tool. In doing so. the Commission has imposed upon itself a particular management process as a legal requirement which cannot be ignored. adapted to circumstances. or changed without once again going through formal rulemaking procedures. The rule provides a myriad of opportunities for licensees to invoke procedural irregularities in challenging the Commission's rules and its backfitting decisions. The rule lends itself readily to being used as a delaying tactic by uncooperative licensees. and it has the potential for hamstringing the Commission's ability to impose needed safety improvements while the legal wrangling goes on and on and on.

The Commission's decision to include Commission rulemaking within the coverage of the backfit rule is an excellent example of the overproceduralization of the backfit process. The Administrative Procedure Act and cases interpreting it set out requirements for rulemaking. Interested parties are given an opportunity to comment on the proposed agency action before it goes into effect. The Commission must then take those comments into account before promulgating a final rule. The courts can then review Commission action and test it for reasonableness and rationality. The Commission wishes to add on to these legal requirements a very high standard or threshold the Commission must meet before it can institute safety improvements. Further. the rule requires a strict cost-benefit balance. something the courts have not found is required by the Atomic Energy Act or the Administrative Procedure Act. Contrary to Chairman Palladino's assertion. the

Commission cannot decide whether to follow these new. more stringent requirements in individual rulemaking proceedings on a case-by-case basis. This regulation now becomes binding on the Commission. and must be followed in all future rulemakings.

In addition to this. the backfit rule applies to staff interpretations of Commission rules. By the rule's literal terms. any staff interpretation of a Commission rule would also have to meet the requirements of the backfit rule. Thus. the Commission would be required to meet a high threshold and perform a cost-benefit balance for any rule it issues. and the staff would then have to again meet the same high threshold and perform a new cost-benefit balance before it could interpret that rule. That is absurd. but that is what the rule appears to require.

Even if rulemaking were not to be included within the scope of this rule. staff interpretations of Commission rules would be. This presents an interesting dilemma. The Commission's 1985 Policy and Planning Guidance states: "The Commission intends to shift its regulatory emphasis away from detailed. prescriptive requirements toward performance criteria." See. NUREG–0885 Issue 4. "U.S. Nuclear Regulatory Commission Policy and Planning Guidance 1985." This means that new rules will tend to be general in nature. It is difficult for safety reviewers and inspectors to review and inspect generalities. They need to develop positions on acceptance criteria. hardware requirements. applicable quality assurance provisions. technical specifications. etc. The rule would require that these interpretations meet the requirements of the backfit rule. Thus. if rulemakings are outside the scope of the backfit rule. but interpretations of those rules are not. it may create a situation where the staff cannot adequately interpret the rule because the interpretations would not meet the backfitting requirements. even though the Commission's rule has been otherwise legally promulgated.

If these are the results the Commission intended. then the Commission's backfit rule makes no sense whatsoever. If this is not what the Commission intended. then the Commission should make that absolutely clear. Such ambiguities do nothing but provide fodder for litigation. These problems illustrate further how poorly written and how poorly explained is this rule. These are the kinds of issues one should not have to make guesses about.

It is interesting to note that my colleagues constantly complain about the length and over-legalization of the licensing process, often arguing that there are too many procedures and that the process is too formal. Yet, those same Commissioners who want to make the licensing process as simple and informal as possible have here added layer upon layer of procedures to the backfitting process. In fact, they have added procedures far beyond those which are legally required, and in the process have added new opportunities for litigation.

Barrier to Improved Safety

The Commission's rule sets up a threshold standard that the Commission must meet before it can adopt new safety requirements. Under this rule, the Commission cannot reduce the radiological risks to the public unless it first determines that a proposed safety improvement provides a "substantial increase in the overall protection of the public health and safety or the common defense and security." Section 50.109(a)(3). Thus, the Commission creates a significant barrier to reactor safety improvements.

The Commission's explanaton in the statement of considerations of what it means by "substantial increase" is so unclear as to be useless. However, an indication of what the Commission really intends can be found in the Commission's recent Indian Point decision. 21 NRC 1043. In that proceeding. the Commission's technical staff and Licensing Board urged the Commission to require a set of safety improvements for the reactors sited at the most highly populated locations in this country. Upon learning that those improvements only cut the risks to the public in half, the Commission rejected them as not providing a "substantial" increase in protection. 21 NRC 1043.

I would prefer a standard which does not set such a high threshold for the imposition of safety improvements. In fact, I proposed such a standard. Under my proposed standard the Agency would require improved safety upon a determination that a proposed measure provides a net increase in the protection of the public health and safety and that the costs of this improvement are not incommensurate with the increased protection. This standard would allow more improvements in safety than would the Commission's standard but would still exclude proposed changes which would result in only trivial safety improvements. The Commission rejected this standard because it does not present a high enough barrier to block new safety requirements. Apparently,

the Commission is interested in more than simply bringing discipline to the backfitting process. Rather, it is really interested in tying its own and the staff's hands to restrict the number of safety improvements. If the Commission really only wanted discipline and sound thinking to be brought to the backfitting process, it would not feel the need to propose such a stringent threshold for safety improvements.

Tipping the Scale Against Safety

The rule specifies nine factors that "are to be used as balancing mechanisms in the decisionmaking process for backfitting". See, § 50.109(c) for list of factors. If one cuts through the extraneous matter in that section of the rule, one finds that the Commission requires cost-benefit analyses to be performed on all proposals for backfits. Of course, in cost-benefit analyses the bottom line depends on what factors, one chooses to put on the scale.

Not satisfied that a high threshold standard will sufficiently limit the number of backfits, the Commission has also decided to stack the cost-benefit balance. The only benefit the Commission is able to identify as being appropriately considered in decisions on whether safety should be improved is the "potential change in the risk to the public from the accidental off-site release of radioactive material." Section 50.109(c)(3). Risk is typically defined as the probability of an accident multiplied by the consequences, with the latter expressed as the collective dose to the public (person-rem). However, even here the Agency's typical practice is to ignore societal doses beyond a 50-mile radius. As calculations of accident consequences indicate, this procedure captures less than half of the health consequences of core meltdown accidents.

The Commission refuses to include among the "balancing factors" the averted costs of off-site property damage resulting from radiological releases. The Commission does not seem to realize that core meltdown accidents can contaminate off-site property to hazardous radiation levels and that there is a real benefit in preventing that from occurring. Averting the necessity to decontaminate such property is a real benefit of backfits which lessen the likelihood of off-site releases of radioactive materials. Since these costs in some instances substantially exceed the monetized value of averted health effects resulting from accidents, the Commission has no defensible basis for omitting from the "balancing factors" off-site property decontamination costs.

The Commission also rejects the inclusion of the benefits derived from averting damage to the plant itself. The TMI-2 accident, which apparently did not result in extensive melting of the reactor core or substantial offsite releases of radioactivity, resulted in billions of dollars in plant damage, plant clean-up and power replacement costs. The Commission's rule fails to recognize that preventing such costs has a public benefit. The Commission chooses to ignore averted replacement power costs associated with safety improvements that prevent accidents. However, in order to inflate the costs side of the equation which weighs against backfitting, the rule requires consideration of the replacement power costs for the facility downtime associated with implementing a backfit.

At the same time the Commission ignores the benefits of backfits, the Commission tries to include every conceivable "cost" of the backfit in the balance. The rule includes costs such as installation and other costs associated with physically changing the plant: the cost of facility downtime, e.g. replacement power costs, etc: the cost of construction delay; and, radiological impact on facility employees. The Commission has even thrown the cost to the NRC (resource burden on the NRC) into the balance. Obviously this stacking of the deck against safety improvements indicates once again that the Commission is interested in more than just adding discipline to the backfitting process.

The Commission majority tries to argue that the balance of costs and benefits is not slanted because other benefits beyond those enumerated in the rule can be considered. Their own actions contradict this argument. In adopting this rule, the Commission majority expressly rejected proposals to include additional health and safety and economic benefits of proposed backfits that would have resulted in a fair and even-handed consideration of all relevant costs and benefits. Given its own actions, the true intent of the Commission majority is beyond doubt.

Reliance on Indefensible Analyses

The Commission's rule places great reliance on Probabilistic Risk Analyses (PRA's). In determining the "change in risk" as required by this rule, the Commission intends to rely on the bottom-line results of PRA's. Unfortunately, numbers produced by these analyses amount really to only estimated guesses: yet, the Commission intends to rely heavily on these numbers, which nearly all PRA practitioners agree are unreliable. in determining whether to require improvements in safety.

Preparation of a PRA requires that the analyst calculate the core meltdown probability. Given a particular core meltdown scenario. the analyst must then calculate the containment failure mode. the quantity of radioactive fission products released from the containment (the "source term"). the dispersion of the fission products in the atmosphere and finally the radiation doses to the public. The calculated probabilities from all of the above are multiplied by the aggregate doses to the public. This is the risk to the public.

To calculate the change in risk. as required by this rule. the analyst must first calculate the risk to the public before a proposed safety improvement is implemented. and then calculate the risk assuming the improvement is made. Unfortunately the necessary calculations cannot be made based on data. and scientifically accepted principles and methodologies. Because of major inadequacies in the data base. because of the vast complexity of nuclear plants. because a tremendous number of assumptions must be made. because all contributors to risk cannot be quantified and because core meltdown phenomena are poorly understood. no one calculation of risk yields a remotely meaningful value of risk. I discussed the meaningless nature of these risk estimates in more detail in my separate views on the Severe Accident Policy Statement.

Our experience with the Davis Besse plant provides an excellent example of the inadequacies of PRA's for truely predicting risk. It also illustrates the shortcomings of a system which relies heavily on strict cost-benefit balances for making decisions on safety improvements. The Davis Besse plant has one of the most (if not the most) unreliable emergency feedwater systems (EFS) of any nuclear plant in this country. The NRC staff has been trying to require Davis Besse to upgrade its EFS reliability. However. for the last several years. the licensee has been using reliability and cost-benefit analyses to argue that substantial upgrades should not be required. Two independent reliability analyses (one by the utility and one by the NRC staff) were performed on the EFS at Davis Besse. The results of these two studies differed by a factor of 100 in their estimate of the reliability of the systems. The studies also differed on what was the most cost-effective way to upgrade the system. The utility argued that its cost-benefit analyses showed that only

some low-cost minor changes were justifiable while the staff argued that its cost-benefit analyses supported more significant modifications. Because the Commission-required cost-benefit analyses could not demonstrate the necessity of a particular way to upgrade the EFS reliability. the staff could not require a substantial upgrade of that system even though the plant continued to operate with an unreliable but crucial safety system during the several years of the PRA debate between the staff and the utility.

The June 9. 1985 Davis Besse event demonstrated that the PRA analyses were wrong. Davis Besse had a loss of all feedwater that involved the failure of 14 separate pieces of equipment. (See. NUREG-1154. "Loss of Main and Auxiliary Feedwater Event at the Davis-Besse Plant on June 9. 1985"). The event led the Agency's chief safety officer to observe: "I believe that the recent Davis-Besse event illustrates that. in the real world. system and component reliabilities can degrade below those we and the industry routinely assume in estimating core melt frequencies." (See memorandum from Harold R. Denton to William J. Dircks. dated June 27. 1985). Further. it appears that the steam and feedwater rupture control system had a significant role in causing the loss of emergency feedwater. Yet that system was not even included as a possible contributor to the unreliability of the emergency feedwater system in either of the independent reliability studies. Despite this clear evidence of the weaknesses in PRA studies and their potential for manipulation and distortion. the Commission persists in using them and in requiring their use by the staff in this rule as the basis for deciding on safety improvements.

Although this rule will have a negative impact on all aspects of the Commission's reactor safety activities. its effects are likely to be greatest in the area of improving human performance. Recent operating experience indicates that roughly half of all significant operating events can be traced to inadequate human performance in such areas as reactor operations. surveillance testing and maintenance. A number of the Commission's post-TMI requirements have focused on human performance. but recent operating experience demonstrates the need for further improvements in this area. Indeed. virtually all members of the Commission have advocated further measures to improve the qualifications. experience and training of plant personnel. Specifically. members of the Commission have spoken in favor of

increasing the engineering knowledge and skills of plant operators and requiring the use of plant-reference simulators for operator training and testing. Common sense and sound engineering judgment tell us that such measures will have a positive effect in improving plant performance. Yet. it will be especially difficult to assess how such proposed requirements will reduce the risk of a core melt accident which might result in harm to the public. Thus. the practical effect of this rule will be to thwart the efforts of the NRC staff to develop new safety requirements in the area of human performance where such requirements could be of the greatest safety benefit.

Ignoring Uncertainties

The Commission also fails to deal with the huge uncertainties associated with the risks of nuclear reactors. The actual risks could be up to 100 times the value frequently picked by the Commission. One would think that the uncertainties about the level of safety achieved at the operating reactors would have a bearing on whether reactor safety should be improved. I proposed that the Commission articulate its expectations on the handling of uncertainties in the backfitting decisionmaking process before allowing this rule to become effective. The Commission rejected my proposal. There is no reference in this rule to uncertainties in reactor risks or to how uncertainties are to be factored into safety decisions. The Commission's silence simply reaffirms its practice of ignoring the enormous uncertainties in nuclear risks when deciding whether to improve the protection of the public health and safety.

Selective Application of the Rule

The Commission's stated manner of applying this rule is also troubling. First. according to the statement of Considerations. a licensee may request an amendment to its license and the NRC staff is not required to consider whether the amendment represents a "substantial increase in the overall protection of the public health and safety." However. if the NRC staff wants to amend a license to establish a more stringent standard. the staff must first demonstrate that the amendment meets that backfitting standard. Thus. the rule stacks the deck in favor of the industry and against the NRC staff.

But more troubling is the Commission's apparent intent to apply this backfit rule with its high threshold and cost-benefit analysis only to those new Commission requirements which are intended to improve safety. If one reads the Commission's rule literally, it applies to *any* change in Commission requirements. both a change to make requirements more stringent and one to relax requirements. Further, the Commission states in its statement of considerations: "(T)here is no intent on the part of the Commission to include within the scope of the exceptions (to the rule) new or modified interpretations of what constitutes no undue risk to the public health and safety. In such a case. the rule applies." All of this seems to indicate that the backfit rule applies across the board to new Commission regulations and interpretations.

However, the Commission's actions and rhetoric would seem to indicate otherwise. The Commission has been devoting and continues to devote considerable agency resources to relaxing the current emergency core cooling regulations and the emergency planning regulations. For example, the staff is developing new and relaxed (relative to the current staff position) acceptance criteria for emergency core cooling systems that would effectively allow the licensees to increase the power level of the operating reactors. Likewise, the Commission assigns resources to work on a rule that would allow less comprehensive evacuation planning. Both activities involve new or modified interpretations of what constitutes compliance. involve a modified interpretation of what constitutes "no undue risk." and do not fall within any of the exemptions to the backfit rule. Thus, if one reads the backfit rule literally, the Commission must determine that increasing the power level of reactors and diminishing the level of emergency preparedness result in a "substantial increase in the overall protection of the public health and safety or common defense and security." It would take quite a bit of convoluted argument to find that relaxing safety standards meets the rule's substantial increase requirement. One can only conclude that either the Commission is wasting resources on these activities or that it does not intend to apply the backfit rule to actions which *relax* existing safety standards.

This problem of interpretation is another example of how poorly thought out and how poorly written is this rule. The Commission should make clear exactly what is the scope of this rule, and should revise the rule accordingly. Otherwise. this apparent ambiguity once again produces nothing but fuel for litigation.

Conclusion

I might be as sanguine as is the Commission about the current state of reactor safety, and I might be willing to restrict the Commission's ability to require safety improvements if there were a clear understanding of the level of safety already achieved at plants and if that understanding demonstrated that the potential for severe accidents is indeed very remote. Unfortunately. the Commission does not have a clear understanding of the level of safety of current reactors.

The Commission does not know where we are on the learning curve for reactor performance, and there is a distinct possibility of one or more severe accidents in the foreseeable future. Operating experience indicates that a total loss of a safety system is not a rare event, that multiple independent failures do occur, that there are component and reliability problems. that operating practices are frequently deficient. and that there are a wide range of adverse systems interactions. The Commission is reluctant to face these facts and to demand improved safety because that might suggest to the public that the existing reactors are unsafe and might hinder the further development of the nuclear industry.

In my view, another severe accident may well bring to a halt further development of the nuclear industry and. if people are injured, may jeopardize continued operation of existing reactors. The Commission has said there is about a 50–50 chance of another severe accident in the next twenty years. The Commission finds that risk so acceptable that it can now, through this rule, put roadblocks in the way of further safety improvements. I find the Commission's actions to be not only unwise but harmful to the public interest and potentially hazardous to the public health and safety.

The Commission will next turn its attention to the forth and final action that will complete the framework for deciding whether the NRC and the industry will pursue safety issues before accidents occur, i.e. the Safety Goal Policy Statement. That will be the final opportunity to come to grips with the pivotal issues the Commission has steadfastly avoided over the last several years. As I wrote in my separate views on the Severe Accident Policy Statement. It is encouraging that there appears to be an emerging consensus within the NRC senior technical staff and within the ACRS in favor of safety improvements to reduce severe accident risks. However, it is dismaying that the Commission. having lost all sight of the

broadest lessons learned from the TMI-2 accident. has chosen to hinder enhancing the protection of the public health and safety through this backfit rule.

Views of Commissioner Bernthal

I had fully expected to support the Commission's final rule on backfitting. Unfortunately, an eleventh-hour decision by the majority has added a destructive provision that at best can only confuse the public and our licensees by its misrepresentation of the role and options of the Commission in rulemaking; at worst it contains the seeds for rulemaking chaos, with litigative risks. unpredictability, and lengthened timetables that will result in more. rather than less uncertainty in the Commission's entire licensing and regulatory process. Such a backfitting rule is surely not in the public interest or in the interest of our licensees.

In a word. my principal quarrel with the rule adopted by the Commission is its inclusion of rulemaking in the definition of backfitting. Indeed, the mere idea of imposing its own rule on the statutory procedures for rulemaking as set forth in the Administrative Procedures Act should have given the Commission majority long pause. to say the least.

But in its apparent desire to appear to have voluntarily circumscribed its own authority and flexibility for rulemaking (when it cannot, of course, ultimately do so). the Commission has instead chosen to run the risk of creating new. legally binding requirements for rulemaking. requirements which will only widen the target for anyone seeking to challenge a final rule.

It is not even clear just who it is the Commission believes will be served by this action. Far from lending discipline and order to the rulemaking process, what the Commission majority has done will help insure that our often long and tortured consideration of rules will become even longer, more tortured. and more confusing. More ominously, should a future Commission find common-sense public health and safety measures unduly confused and obstructed by the backfit rule. it may in frustration choose simply to begin issuing by order "rules" that today would be subjected to the careful. disciplined process set forth in the Administrative Procedures Act.

The only rationale the majority has offered for wanting to include rulemaking under the backfit rule is to "discipline" the Commission (i.e. to protect the Commission from itself). If the Commission is incapable of disciplining itself in the rulemaking process as it stands (what with the existing Committee to Review Generic Requirements and the Commission's incontestable authority and ineluctable responsibility to instruct the staff). then I doubt that rule laid upon rule will do much to teach the Commission the virtue of self-discipline.

More specifically. the Commission majority presumably knows that the backfit threshold criteria applied to rulemaking would apply not just on a plant-specific basis (which it should be recalled was the intent of the original backfitting initiative). but to *generic* decisions that may affect dozens of plants, and in fact to rulemaking on *all* but procedural matters. rulemaking that may or may not have the remotest connection to what the public and our licensees normally consider a plant "backfit". The scope of Commission rulemaking responsibilities thus often involves broad public policy considerations. and those considerations can rise above elements as simple as cost-benefit analysis to reach issues as fundamental as fairness and individual rights. The Commission's backfit rule. if applied to rulemaking itself. will thus serve only to trivialize in appearance and confuse in practice the many factors to be weighed in rulemaking.

As one small example of the morass into which the Commission majority has wandered. consider (as the Commission currently is considering) whether there should be a requirement that radiation workers be provided their dose records annually. The "benefit" of this "backfit" of Commission rules may seem clear. but it might very well never pass the cost-benefit test. Indeed. it is difficult to imagine a rule that would involve the human-factors element of plant operations. and that would also be amenable to straightforward cost-benefit analysis.

Rulemaking as it exists involves numerous inherent procedural checks and balances to insure that each proposal is carefully considered prior to adoption. Indeed. rulemaking is the forum which provides the greatest number of checks against arbitrary action by the Staff or Commission. Much of the analysis (*including* cost-benefit) which the new backfitting rule would require is already done informally throughout the process of considering and adopting new regulations.

If the Commission wishes to insure still more structure in the rulemaking process. structure which could take into account every single factor set forth in the backfit rule and more. there are ample means of doing so by simple internal agency management. Such

methods would reaffirm existing Commission guidelines to the Staff without opening the door to additional needless litigation as a consequence of vague new. legally enforceable. Commission-created rights added to those already available to all parties under the APA.

The entire backfit rulemaking was undertaken to bring order and accountability to plant modifications heretofore sometimes imposed without the benefit of systematic evaluation and justification. In rulemaking per se. that objective has always been well within the Commission's grasp—it is. after all. the Commission that makes rules. For good measure. the Commission also has the Administrative Procedures Act as a matter of law. and its own Committee to Review Generic Requirements as a matter of internal administrative policy to assist it in carrying out such considered decision-making. Casting the net of the new backfit rule over Commission rule-making (almost as an afterthought. as it happened in this case) is thus at best an exercise in pointless symbolism. and at worst potentially destructive of the Commission's entire rule-making process.

Unneeded law is bad law. and unneeded regulation is bad regulation. The Commission majority has imposed on this agency new regulatory obligations in rulemaking that are not only unneeded. but which the Commission majority itself hopes and trusts will be of little practical (i.e. legally enforceable) consequence. To the extent that this rule will affect rulemaking. it will therefore be a bad rule. In sum. the Commission majority has inexplicably insisted on fixing not only what is. but what ain't broke. I will not be a party to such poor judgment.

Environmental Impact: Categorical Exclusion

The NRC has determined that this final rule is the type of action described in categorical exclusion 10 CFR 51.22(c)(3). Therefore. neither an environmental impact statement nor an environmental assessment has been prepared for this final rule.

Paperwork Reduction Act Statement

This final rule does not contain a new or amended information collection requirement subject to the Paperwork Reduction Act of 1980 (44 U.S.C. 3501 *et seq.*). Existing requirements were approved by the Office of Management and Budget. Approval Number 3150–0011.

Regulatory Flexibility Act Certification

In accordance with the Regulatory Flexibility Act of 1980. 5 U.S.C. 605(b). the Commission hereby certifies that this final rule. if promulgated. will not have a significant economic impact on a substantial number of small entities. The affected facilities are licensed under the provisions of 10 CFR 50.21(b) and 10 CFR 50.22. The companies that own these facilities do not fall within the scope of "small entities" as set forth in the Regulatory Flexibility Act or the Small Business Size Standards set forth in regulations issued by the Small Business Administration in 13 CFR Part 121.

List of Subjects

10 CFR Part 2

Administrative practice and procedure. Nuclear power plants and reactors. hazardous waste.

10 CFR Part 50

Antitrust. Classified information. Fire prevention. Incorporation by reference. Intergovernmental relations. Nuclear power plants and reactors. Penalty. Radiation protection. Reactor siting criteria. Reporting and recordkeeping requirements.

For the reasons set out in the preamble and under the authority of the Atomic Energy Act of 1954. as amended. the Energy Reorganization Act of 1974. as amended. and 5 U.S.C. 553. the NRC is adopting the following amendments to 10 CFR Parts 2 and 50.

PART 50—DOMESTIC LICENSING OF PRODUCTION AND UTILIZATION FACILITIES

1. The authority citation for Part 50 continues to read as follows:

Authority: Secs. 103. 104. 161. 182. 183. 185. 189. 68 Stat. 936. 937. 948. 953. 954. 955. 956. as amended. sec. 234 83 Stat. 1244. as amended (42 U.S.C. 2133. 2134. 2201. 2232. 2233. 2236. 2239. 2282); secs. 201. 202. 206. 68 Stat. 1242. 1244. 1246. as amended (42 U.S.C. 5841. 5842. 5846). unless otherwise noted.

Sec. 50.7 also issued under Pub. L. 95–601. sec. 10. 92 Stat. 2951 (42 U.S.C. 5851). Sections 50.57(d). 40.58. 50.91. and 50.92 also issued under Pub. L. 97–415. 96 Stat. 2071. 2073 (42 U.S.C. 2133. 2239). Section 50.78 also issued under sec. 122. 68 Stat. 939 (42 U.S.C. 2152). Sections 50.80–50.81 also issued under sec. 184. 68 Stat. 954. as amended (42 U.S.C. 2234). Sections 50.100–50.102 also issued under sec. 186. 68 Stat. 955 (42 U.S.C. 2236).

For the purposes of sec. 223. 68 Stat. 958. as amended (42 U.S.C. 2273). §§ 50.10 (a). (b). and (c). 50.44. 50.46. 50.48. 50.54. and 50.80(a) are issued under sec. 161b. 68 Stat. 948. as amended (42 U.S.C. 2201(b)): §§ 50.10 (b) and (c) and 50.54 are issued under sec. 161i. 68

Stat. 949, as amended (42 U.S.C. 2201(i)); and §§ 50.53(e), 50.59(b), 50.70, 50.71, 50.72, 50.73, and 50.78 are issued under sec. 161o. 68 Stat. 950, as amended (42 U.S.C. 2201(o)).

2. In § 50.54, paragraph (f) is revised to read as follows:

§ 50.54 Conditions of licenses.

*   *   *

(f) The licensee shall at any time efore expiration of the license, upon :quest of the Commission submit written statements, signed under oath or affirmation, to enable the Commission to determine whether or not the license should be modified, suspended or revoked. Except for information sought to verify licensee compliance with the current licensing basis for that facility, the NRC must prepare the reason or reasons for each information request prior to issuance to ensure that the burden to be imposed on respondents is justified in view of the potential safety significance of the issue to be addressed in the requested information. Each such justification provided for an evaluation performed by the NRC staff must be approved by the Executive Director for Operations or his or her designee prior to issuance of the request.

*   *   *   *   *

3. In § 50.109, paragraph (a) is revised, paragraph (b) is removed, paragraph (c) is revised and redesignated as (b), and new paragraphs (c), (d) and (e) are added to read as follows:

§ 50.109 Backfitting.

(a)(1) Backfitting is defined as the modification of or addition to systems, structures, components, or design of a facility; or the design approval or manufacturing license for a facility; or the procedures or organization required to design, construct or operate a facility; any of which may result from a new or amended provision in the Commission rules or the imposition of a regulatory position interpreting the Commission rules that is either new or different from a previously applicable staff position after:

(i) The date of issuance of the construction permit for the facility for facilities having construction permits issued after October 21, 1985; or

(ii) Six months before the date of docketing of the operating license application for the facility for facilities having construction permits issued before October 21, 1985; or

(iii) The data of issuance of the operating license for the facility for facilities having operating licenses; or

(iv) The date of issuance of the design approval under Appendix M, N or O of this part.

(2) The Commission shall require a systematic and documented analysis pursuant to paragraph (c) of this section for backfits which it seeks to impose. Imposition of a backfit pursuant to paragraph (a)(4)(ii) of this section shall not relieve the Commission of performing an analysis after the fact to document the safety significance and appropriateness of the action taken.

(3) The Commission shall require the backfitting of a facility only when it determines, based on the analysis described in paragraph (c) of this section, that there is a substantial increase in the overall protection of the public health and safety or the common defense and security to be derived from the backfit and that the direct and indirect costs of implementation for that facility are justified in view of this increased protection.

(4) The provisions of paragraphs (a)(2) and (a)(3) of this section are inapplicable and, therefore, backfit analysis is not required and the standard does not apply where the staff finds and declares, with appropriate documented evaluation for its finding, either:

(i) That a modification is necessary to bring a facility into compliance with a license or the rules or orders of the Commission, or into conformance with written commitments by the licensee; or

(ii) That an immediately effective regulatory action is necessary to ensure that the facility poses no undue risk to the public health and safety.[1]

Such documented evaluation shall include a statement of the objectives of and reasons for the modification and the basis for invoking the exception.

(b) Paragraph (a) of this section shall not apply to backfits imposed prior to October 21, 1985.

(c) In reaching the determination required by paragraph (a) of this section, the Commission will consider how the backfit should beiprioritized and scheduled in light of other regulatory activities ongoing at the facility and, in addition, will consider information available concerning any of the following factors as may be appropriate and any other information relevant and material to the proposed backfit:

(1) Statement of the specific objectives that the proposed backfit is designed to achieve;

(2) General description of the activity that would be required by the licensee or applicant in order to complete the backfit;

(3) Potential change in the risk to the public from the accidental off-site release of radioactive material;

(4) Potential impact on radiological exposure of facility employees;

(5) Installation and continuing costs associated with the backfit, including the cost of facility downtime or the cost of construction delay;

(6) The potential safety impact of changes in plant or operational complexity, including the relationship to proposed and existing regulatory requirements;

(7) The estimated resource burden on the NRC associated with the proposed backfit and the availability of such resources;

(8) The potential impact of differences in facility type, design or age on the relevancy and practicality of the proposed backfit;

(9) Whether the proposed backfit is interim or final and, if interim, the justification for imposing the proposed backfit on an interim basis.

(d) No licensing action will be withheld during the pendency of backfit analyses required by the Commission's rules.

(e) The Executive Director for Operations shall be responsible for implementation of this section and all analyses required by this section shall be approved by the Executive Director for Operations or his designee.

4. In Appendix 0 to 10 CFR Part 50, a new section (8) is added to read as follows:

Appendix 0—Standardization of Design; Staff Review of Standard Designs

*   *   *   *   *

8. Information requests to the approval holder regarding an approved design shall be evaluated prior to issuance to ensure that the burden to be imposed on respondents is justified in view of the potential safety significance of the issue to be addressed in the requested information. Each such evaluation performed by the NRC staff shall be in accordance with 10 CFR 50.54(f) and shall be approved by the Executive Director for Operations or his or her designee prior to issuance of the request.

PART 2—[AMENDED]

5. The authority citation for Part 2 continues to read as follows:

Authority: Secs. 161, 181, 68 Stat. 948, 953, as amended (42 U.S.C. 2201, 2231); sec. 191, as amended. Pub. L. 84-615, 76 Stat. 408 (42

---

[1] For those modifications which are to ensure that the facility poses no undue risk to the public health and safety and which are not deemed to require immediately effective regulatory action, analyses are required; these analyses, however, should not involve cost considerations except only insofar as cost contributes to selecting the solution among various acceptable alternatives to ensuring no undue risk to public health and safety.

U.S.C. 2241); sec. 201. 88 Stat. 1342. as amended (42 U.S.C. 5841); 5 U.S.C. 552.

Sec. 2.101 as issued under secs. 52.62, 63.81. 103, 104, 105, 68 Stat. 930, 932, 933, 935, 936. 937, 938, as amended (42 U.S.C. 2073, 2092, 2093, 2111, 2133, 2134, 2135); sec. 102. Pub. L. 91-190. 63 Stat. 853. as amended (42 U.S.C. 4332); sec. 301. 88 Stat. 1248 (42 U.S.C. 5871). Sections 2.102, 2.103, 2.104, 2.105, 2.721 also issued under secs. 102, 103, 104, 105, 183, 189. 68 Stat. 936, 937, 938, 954, 955, as amended (42 U.S.C. 2132, 2133, 2134, 2135, 2233, 2239). Section 2.105 also issued under Pub. L. 97-415, 96 Stat. 2073 (42 U.S.C. 2239). Sections 2.200-2.206 also issued under secs. 186, 234. 68 Stat. 955, 83 Stat. 444, as amended (42 U.S.C. 2236, 2282); sec. 206. 88 Stat. 1246 (42 U.S.C. 5846). Sections 2.300-2.309 also issued under Pub. L. 97-415. 96 Stat. 2071 (42 U.S.C. 2133). Sections 2.600-2.606 also issued under sec. 102. Pub. L. 91-190. 83 Stat. 853 as amended (42 U.S.C. 4332). Sections 2.700a. 2.781 also issued under 5 U.S.C. 554. Sections 2.754, 2.760, 2.770, 2.780 also issued under 5 U.S.C. 557. Section 2.790 also issued under sec. 103. 68 Stat. 936. as amended (42 U.S.C. 2133) and 5 U.S.C. 552. Sections 2.800 and 2.808 also issued under 5 U.S.C. 553. Section 2.809 also issued under 5 U.S.C. 553 and sec. 29, Pub. L. 85-256. 71 Stat. 579. as amended (42 U.S.C. 2039). Appendix A also issued under sec. 6. Pub. L. 91-580. 84 Stat. 1473 (42 U.S.C. 2135).

6. Section 2.204 is revised to read as follows:

§ 2.204   Order for modification of license.

The Commission may modify a license by issuing an amendment on notice to the licensee that the licensee may demand a hearing with respect to all or any part of the amendment within twenty (20) days from the date of the notice or such longer period as the notice may provide. The amendment will become effective on the expiration of the 20-day period during which the licensee may demand a hearing. If the licensee requests a hearing during this 20-day period, the amendment will become effective on the date specified in an order made following the hearing. When the Commission finds that the public health, safety, or interest so requires, the order may be made immediately effective. If the amendment involves a backfit, the provisions of § 50.109 of this chapter shall be followed.

Dated at Washington. D.C.. this 17th day of September, 1985.

For the Nuclear Regulatory Commission.

Samuel J. Chilk.

Secretary of the Commission.

[FR Doc. 85-22572 Filed 9-19-85: 8:45 am]

BILLING CODE 7590-01-M

DEPARTMENT OF HEALTH AND HUMAN SERVICES

Social Security Administration

20 CFR Part 404

[Regulation No. 4]

Federal Old-Age, Survivors, and Disability Insurance; Listing of Impairments—Mental Disorders

Correction

In FR Doc. 85-20552 beginning on page 35038 in the issue of Wednesday, August 28, 1985, make the following corrections:

1. On page 35040, third column. seventh line from the bottom, "of" should read "or".

2. On page 35044, first column. in the fourth *Comment*, sixth line, insert the word "only" between "if" and "one".

3. On page 35045, third column. in the third *Comment*, sixth line, "by" should read "be".

4. On page 35046, first column. in the fourth *Comment*, second line from the bottom, "patient's" should read "patients".

5. On page 35048, first column. in the second *Comment*, first line, "larger" should read "large".

6. On page 35049, first column. in the first *Response*, twelfth line, "necessary" should read "necessarily".

7. On the same page, second column, in the first *Response*, second line from the bottom, "individual" should read "individuals". Also in the third column, "12.04 *Mental Retardation*" should read "12.05 *Mental Retardation*".

8. On page 35066, third column, first complete paragraph, "including" should read "include".

BILLING CODE 1506-01-M

20 CFR Part 404

Social Security Benefits; Coverage of Employees of Private Nonprofit Organizations, Work Outside United States, Etc.

Correction

In FR Doc. 85-21121, beginning page 36571 in the issue of Monday, September 9, 1985, make the following corrections:

On page 36572, first column. in the DATES paragraph:

1. In the first and second lines, "(insert date of publication)" should have read "September 9, 1985";

2. In the eleventh line "received" should have read "receive".

BILLING CODE 1506-01-M

Food and Drug Administration

21 CFR Part 520

Oral Dosage Form New Animal Drugs Not Subject To Certification; Flunixin Meglumine Paste

AGENCY: Food and Drug Administration.

ACTION: Final rule.

SUMMARY: The Food and Drug Administration (FDA) is amending the animal drug regulations to reflect approval of a new animal drug application (NADA) filed by Schering Corp., providing for flunixin meglumine paste. The paste is for oral use in horses to alleviate inflammation and pain from musculoskeletal disorders.

EFFECTIVE DATE: September 20, 1985.

FOR FURTHER INFORMATION CONTACT: Sandra K. Woods, Center for Veterinary Medicine (HFV-114), Food and Drug Administration, 5600 Fishers Lane, Rockville, MD 20857, 301-443-3420.

SUPPLEMENTARY INFORMATION: The Schering Corp., Galloping Hill Rd., Kenilworth, NJ 07033, has filed NADA 137-409 for Banamine® Paste (flunixin meglumine). Flunixin meglumine paste is for the alleviation of inflammation and pain associated with musculoskeletal disorders in horses. The NADA is approved and the regulations are amended to reflect the approval. The basis for approval is discussed in the freedom of information summary.

In accordance with the freedom of information provisions of Part 20 (21 CFR Part 20) and § 514.11(e)(2)(ii) (21 CFR 514.11(e)(2)(ii)), a summary of safety and effectiveness data and information submitted to support approval of this application may be seen in the Dockets Management Branch (HFA-305), Food and Drug Administration, Rm. 4-62, 5600 Fishers Lane, Rockville, MD 20857, from 9 a.m. to 4 p.m. Monday through Friday.

The agency has determined under 21 CFR 25.24(d)(1)(iii) (April 26, 1985; 50 FR 16636) that this action is of a type that does not individually or cumulatively have a significant effect on the human environment. Therefore, neither an environmental assessment nor an environmental impact statement is required.

List of Subjects in 21 CFR Part 520

Animal drugs, oral use.

Therefore, under the Federal Food, Drug, and Cosmetic Act and under authority delegated to the Commissioner of Food and Drugs and redelegated to the Center for Veterinary Medicine, Part 520 is amended as follows:

APPENDIX B

THE 1988 BACKFIT RULE

TABLE I—Continued

Avocado variety	Effective period		Minimum size	
	From	Through	Weight (ounces)	Diameter (inches)
	3rd Mon Sept	1st Sun Oct	12	3-6/16
Guatemalan Seedling [2]	1st Mon Sept	1st Sun Oct	15	
	1st Mon Oct	1st Sun Dec	13	
Marcus	1st Mon Sept	3rd Sun Sept	32	4-12/16
	3rd Mon Sept	5th Sun Oct	24	4-6/16
Brooks 1978	1st Mon Sept	2nd Sun Sept	12	3-4/16
	2nd Mon Sept	3rd Sun Sept	10	3-1/16
	3rd Mon Sept	2nd Sun Oct	8	2-14/16
Collinson	2nd Mon Sept	2nd Sun Oct	16	3-10/16
Rue	2nd Mon Sept	3rd Sun Sept	30	4-3/16
	3rd Mon Sept	1st Sun Oct	24	3-15/16
	1st Mon Oct	3rd Sun Oct	16	3-8/16
Hickson	2nd Mon Sept	4th Sun Sept	12	3-1/16
	4th Mon Sept	2nd Sun Oct	10	3
Simpson	3rd Mon Sept	2nd Sun Oct	16	3-8/16
Choquette	4th Mon Sept	3rd Sun Oct	28	4-4/16
	3rd Mon Oct	5th Sun Oct	24	4-1/16
	5th Mon Oct	2nd Sun Nov	20	3-14/16
Winslowson	4th Mon Sept	3rd Sun Oct	18	3-14/16
Leona	4th Mon Sept	2nd Sun Oct	16	3-10/16
Hall	4th Mon Sept	2nd Sun Oct	26	3-14/16
	2nd Mon Oct	4th Sun Oct	20	3-9/16
	4th Mon Oct	1st Sun Nov	18	3-8/16
Herman	1st Mon Oct	3rd Sun Oct	16	3-8/16
	3rd Mon Oct	5th Sun Oct	14	3-6/16
Lula	1st Mon Oct	3rd Sun Oct	18	3-11/16
	3rd Mon Oct	5th Sun Oct	14	3-16/16
	5th Mon Oct	2nd Sun Nov	12	3-3/16
Ajax (B-7)	2nd Mon Oct	5th Sun Oct	18	3-14/16
Taylor	2nd Mon Oct	4th Sun Oct	14	3-5/16
	4th Mon Oct	1st Sun Nov	12	3-2/16
Booth 3	2nd Mon Oct	3rd Sun Oct	16	3-8/16
	3rd Mon Oct	5th Sun Oct	14	3-6/16
Linda	5th Mon Oct	3rd Sun Nov	18	3-12/16
Monroe	1st Mon Nov	3rd Sun Nov	26	4-3/16
	3rd Mon Nov	1st Sun Dec	24	4-1/16
	1st Mon Dec	3rd Sun Dec	20	3-14/16
	3rd Mon Dec	1st Sun Jan	16	3-8/16
Booth 1	2nd Mon Nov	4th Sun Nov	16	3-12/16
	4th Mon Nov	2nd Sun Dec	12	3-6/16
Zio (P)	2nd Mon Nov	4th Sun Nov	12	3-1/16
	4th Mon Nov	2nd Sun Dec	10	2-14/16
Wagner	3rd Mon Nov	1st Sun Dec	12	3-5/16
	1st Mon Dec	3rd Sun Dec	10	3-2/16
Brookslate	2nd Mon Dec	3rd Sun Dec	18	3-13/16
	3rd Mon Dec	4th Sun Dec	16	3-10/16
	4th Mon Dec	2nd Sun Jan	14	3-8/16
	2nd Mon Jan	4th Sun Jan	12	3-5/16
	4th Mon Jan	1st Sun Feb	10	
Meya (P)	2nd Mon Dec	4th Sun Dec	13	3-2/16
	4th Mon Dec	2nd Sun Jan	11	3
Reed (CP)	2nd Mon Dec	4th Sun Dec	12	3-4/16
	4th Mon Dec	2nd Sun Jan	10	3-3/16
	2nd Mon Jan	4th Sun Jan	9	3
Buccaneer	5th Mon Oct	4th Sun Nov	13	3-6/16

[1] Avocados of the West Indian type varieties and the West Indian type seedlings not listed elsewhere in Table I.
[2] Avocados of the Guatemalan type varieties, hybrid varieties, and unidentified seedlings not listed elsewhere in Table I.

Dated: May 31, 1988.
Robert C. Keeney,
Deputy Director, Fruit and Vegetable Division.
[FR Doc. 88-12551 Filed 6-3-88; 8:45 am]
BILLING CODE 3410-02-M

NUCLEAR REGULATORY COMMISSION

10 CFR Part 50

Revision of Backfitting Process for Power Reactors

AGENCY: Nuclear Regulatory Commission.

ACTION: Final rule.

SUMMARY: The Nuclear Regulatory Commission is promulgating an amended rule which governs the backfitting of nuclear power plants. This action is necessary in order to have a backfit rule which unambiguously conforms with the August 4, 1987 decision of the U.S. Court of Appeals for the District of Columbia in Union of Concerned Scientists, et al., v. U.S. Nuclear Regulatory Commission. This action is intended to clarify when economic costs may be considered in backfitting nuclear power plants.

EFFECTIVE DATE: July 6, 1988.

FOR FURTHER INFORMATION CONTACT:
Steven F. Crockett, Office of the General Counsel, U.S. Nuclear Regulatory Commission, Washington, DC 20555. Phone: (202) 492-1600.

SUPPLEMENTARY INFORMATION:

Background

On September 20, 1985, after an extensive rulemaking proceeding which included sequential opportunities for public comment on an advanced notice of proposed rulemaking (48 FR 44217; September 28 1983) and a notice of proposed rulemaking (49 FR 47034; November 30, 1984), the Commission adopted final amendments to its rule which governs the backfitting of nuclear power plants, 10 CFR 50.109 (50 FR 38097; September 20, 1985). Backfitting is defined in some detail in the rule, but for purposes of discussion here it means measures which are directed by the Commission or by NRC staff in order to improve the safety of nuclear power reactors, and which reflect a change in a prior Commission or staff position on the safety matter in question.

Judicial review of the amended backfit rule and a related internal NRC Manual chapter which partially implemented it was sought and, on August 4, 1987, the U.S. Court of Appeals for the DC Circuit rendered its decision vacating both the rule and the NRC Manual chapter which implemented the rule in part. UCS v. NRC, 824 F.2d 103. The Court concluded that the rule, when considered along with certain statements in the rule preamble published in the Federal Register, did not speak unambiguously in terms that constrained the Commission from considering economic costs in establishing standards to ensure adequate protection of the public health and safety as dictated by section 182 of the Atomic Energy Act. At the same time, the Court agreed with the Commission that once an adequate level of safety protection had been achieved under section 182, the Commission was fully authorized under section 161i of the Atomic Energy Act to consider and take economic costs into account in ordering further safety improvements. The Court therefore rejected the position of petitioners in the case, Union of Concerned Scientists, that economic costs may never be a factor in safety decisions under the Atomic Energy Act.

Because the Court's opinion regarding the circumstances in which costs may be considered in making safety decisions on nuclear power plants was completely in accord with the Commission's own policy views on this important subject, the Commission

decided not to appeal the decision. Instead, the Commission decided to amend both the rule and the related NRC Manual chapter (Chapter 0514) so that they conform unambiguously to the Court's opinion. On September 10, 1987, the Commission published proposed amendments to the rule (52 FR 34223) and provided for a comment period ending on October 13, 1987.[1] The final rule as set out in this document is substantially the same as the proposed rule (52 FR 34223; September 10, 1987).

In this rulemaking the Commission has adhered to the following safety principle for all of its backfitting decisions. The Atomic Energy Act commands the Commission to ensure that nuclear power plant operation provides adequate protection to the health and safety of the public. In defining, redefining or enforcing this statutory standard of adequate protection, the Commission will not consider economic costs. However, adequate protection is not absolute protection or zero risk. Hence safety improvements beyond the minimum needed for adequate protection are possible. The Commission is empowered under section 161 of the Act to impose additional safety requirements not needed for adequate protection and to consider economic costs in doing so.

The 1985 revision of the backfit rule, which was the subject of the Court's decision, required, with certain exceptions, that backfits be imposed only upon a finding that they provided a substantial increase in the overall protection of the public health and safety or the common defense and security and that the direct and indirect costs of implementation were justified in view of this increased protection. The amended rule, set out in this document, restates the exceptions to this requirement for a finding, so that the rule will clearly be in accord with the safety principle stated above.

[1] In its comments on the proposed amendments, the Union of Concerned Scientists asserts that the Federal Register notice of the proposed amendments was technically defective. UCS argues that since the Court had vacated the entire rule, the Federal Register notice should have proposed enactment of an entire, amended, rule, rather than simply amendments to the vacated rule. In weighing the technical merit of UCS' argument, it should be noted that as of the date of the Federal Register notice, the mandate of the Court had not yet issued and the rule was thus still legally in effect. However, the more important consideration is that the notice clearly revealed the Commission's intent to reissue the backfit rule once it had been conformed to the Court's decision. UCS understood this intent and took the opportunity to resubmit the comments it had submitted during the rulemaking leading up to the 1985 revision of the rule. In any event, the Commission is publishing the entire rule in this document.

Particularly in response to the Court's decision, the rule now provides that if the contemplated backfit involves defining or redefining what level of protection to the public health and safety or common defense and security should be regarded as adequate, neither the rule's "substantial increase" standard, nor its "costs justified" standard, see § 50.109(a)(3), is to be applied. (See § 50.109(a)(4)(iii).) Also in response to the Court's decision, see 824 F.2d at 119, the rule now also explicitly says that the Commission shall always require the backfitting of a facility if it determines that such regulatory action is necessary to ensure that the facility provides adequate protection to the health and safety of the public and is in accord with the common defense and security.

On instruction from the Commission, the NRC staff has amended its Manual chapter on plant-specific backfitting to ensure consistency with the Court's opinion. Copies of the revised chapter are available for public inspection in the Commission's Public Document Room, 1717 H Street NW., Washington, DC 20555.[2]

Response to Comments

Comments were received from 12 utilities, one Federal agency (DOE), one vendor, seven individuals, seven citizens' groups, and two industry groups. Lengthy and detailed comments were submitted by the Union of Concerned Scientists (UCS) and the Nuclear Utility Backfitting and Reform Group (NUBARG). Both organizations were active in the rulemaking which led to the 1985 revision of the rule. The comments submitted by these two groups encompassed most of the comments made by others. Below, the Commission paraphrases the chief comments and responds to them. The Commission has given careful consideration to every comment. The original comments may be viewed in the NRC's Public Document Room in Washington, DC.

[2] Several commenters argue that the revised Manual chapter should undergo what amounts to notice and comment rulemaking. However, the Manual chapter, if it is a rule at all, is a rule of agency organization, procedure, or practice, and therefore is not subject to the notice and comment requirements of the Administrative Procedure Act. See 5 U.S.C. 553(b)(A); see also § 553(a)(2). The Commission did publish for comment an earlier version of Manual Chapter (49 FR 16900; April 20, 1984), but that version was already in effect when it was published for comment, and it was published for comment only because the Commission was still in the process of making fundamental changes to the backfitting process and wanted comment on the procedures then in effect. See id.

"Adequate Protection"

The great majority of the commenters raised issues about the rule's use of the phrase "adequate protection". This phrase is used in the rule's exception provisions. See § 50.109(a)(4). Generally, the rule requires, among other things, that it be shown for a given proposed backfit that implementation of the backfit would bring about a "substantial increase" in overall protection to public health and safety, and that the direct and indirect costs of the backfit are justified by that substantial increase. See § 50.109(a)(3). However, § 50.109(a)(4) also requires that these two standards not be applied in three situations:

First, where the backfit is required to bring a facility into compliance with NRC requirements or the licensee's own written commitments;

Second, where the backfit is necessary to ensure that the facility provides adequate protection to the health and safety of the public and is in accord with the common defense and security; and

Third, as noted above, where the backfit involves defining or redefining what level of protection to the public health and safety or common defense and security should be regarded as adequate.

The comments on the rule's use of the phrase "adequate protection" generally took two forms, each discussed more fully later on in this notice. The first form, most fully represented by UCS' comments, was that the rule itself should actually include a definition of "adequate protection" (the final rule set out in this document does not), a phrase nowhere explicitly defined in general terms, either in the Atomic Energy Act, from which the phrase comes, or in the Commission's regulations.

The second, more modest, form of the comments on "adequate protection", most fully represented by NUBARG's comments, was that one or another of the three exception provisions in the rule was redundant (none is). While not amounting to a call for a definition of "adequate protection", NUBARG's comments displayed some of UCS' uncertainty about what the Commission meant by the phrase.

Each group had difficulty applying the phrase to characterize past Commission action in backfitting. UCS claimed that the Commission had never backfitted in order to achieve something beyond "adequate protection." NUBARG, however, claimed that the Commission had never required a backfit on the grounds that compliance with the regulations was not enough to provide

adequate protection. These views, differing in emphasis, reflect the two groups' opposite concerns about the possibility that the Commission would use the phrase "adequate protection" arbitrarily. UCS is concerned that the Commission might interpret the phrase "adequate protection" to refer to a level of safety such that every proposed improvement would be subjected to cost-benefit analysis. Conversely, the industry appears concerned that the Commission might interpret the phrase "adequate protection" to refer to a level of safety such that no proposed improvement would be subjected to cost-benefit analysis.

The Commission certainly did not intend that this rulemaking should focus on the meaning of the phrase "adequate protection". The main point of this rulemaking was simply to negate the misimpression left by two statements in the preamble to the 1985 version of the backfit rule. UCS puts forward two grounds for its emphasis on the phrase "adequate protection". First, UCS asserts that "(t)he crucial decision as to whether cost benefit analysis will be used in assessing the need for backfitting is dependent on whether the particular backfitting under consideration is needed to ensure adequate safety * * *." Second, UCS claims that the Court "ordered" the Commission to "stop trying to obscure its intentions through ambiguous and vague language * * *."

However, as will be explained more fully below, the Court's decision turned not on the rule's lack of a definition of "adequate protection" but rather on two statements which seemed to the Court to imply that the Commission intended to take costs into consideration in determining what "adequate protection" required; the meaning of "adequate protection" was simply not an issue in the litigation. Moreover, UCS overestimates the role the phrase "adequate protection" plays in the backfit rule. The threshold decision in considering a proposed backfit, and very often the only decision that need be made,[3] is not whether adequate protection is at stake but rather whether the facility is in compliance with Commission's requirements and the licensee's written commitments.

Even if UCS is right about the importance of the phrase "adequate protection", there is nothing unusual or

---

[3] For instance, a majority of the plant-specific backfits carried out during the first year after the 1985 revision of the backfit rule became effective were for the sake of compliance. See SECY-86-46, Evaluation of Managing Plant-Specific Backfit Requirements [November 21, 1986], Enclosure 1.

imprudent, and certainly nothing illegal, about decisions which ultimately turn on the application—by duly constituted authority and after full consideration of all relevant information—of phrases which are not fully defined. Consider, for instance, the "reasonable assurance" determination the Commission must make before issuing an operating license.[4] Indeed, most of the Commission's rules and regulations are ultimately based on unquantified and, as we note below, presently unquantifiable ideas of what constitutes "adequate protection".

Were there something peculiarly critical about the role of "adequate protection" in the backfit rule, the issue of the phrase's meaning could have been raised in the rulemaking for the 1985 rule. Two of the three exception provisions set out above were in the 1985 revision of the rule, where they used the equivalent phrase "undue risk" instead of "adequate protection". Also, as the Court in UCS v. NRC noted, 824 F.2d at 119, the statement of considerations which accompanied the 1985 version of the rule quite explicitly at least twice limited the consideration of costs in backfitting decisions to situations where "adequate protection" was already secured.[5]

Nonetheless, an issue which is a concern of almost every commenter in this rulemaking should not be ignored. Therefore, the Commission will answer as best it can the questions the commenters have raised concerning the rule's use of the phrase "adequate protection". We begin with UCS' call for an objective and generally applicable definition of "adequate protection". We argue that such a definition is not possible in the near future, but that the public and licensees are nonetheless protected against misuse of the phrase. In the course of responding to UCS' comments, we shall, of necessity, be making at least preliminary responses to most of NUBARG's comments also.

UCS argues that the rule permits the agency to escape its legal responsibility

---

[4] "* * * (A)n operating license may be issued by the Commission * * * upon finding that: * * * (t)here is reasonable assurance * * * that the activities authorized by the operating license can be conducted without endangering the health and safety of the public * * *." 10 CFR 50.57(a)(3).

[5] "The consideration and weighing of costs contemplated by the rule applies to backfits that are intended to result in incremental safety improvements for a plant that already provides an acceptable degree of protection(.)" 50 FR 38103, col. 1; also, "(t)he costs associated with proposed new safety requirements may be considered by the Commission provided that the Atomic Energy Act finding 'no undue risk' can be made." Id. at 38101, col. 3.

to articulate the factors on which it bases its backfitting decisions. UCS asserts that the rule should "enunciate criteria and guidelines about what constitutes redefining and defining adequate protection levels, what constitutes an adequate as opposed to a beyond adequate protection level, and what factors place a particular circumstance within the rule or within the exceptions." Another comment asserts that any definition of "adequate protection" should include the resolution of all outstanding safety issues. Yet another calls for "objective criteria", "some real numbers" on releases, accident consequences, and the like.

There does not exist, and cannot exist, at least not yet, a generally applicable definition of "adequate protection" which would guard against every possible misuse of the phrase. Congress established "adequate protection" as the standard the Commission is to apply in licensing a plant, see 42 U.S.C. 2232(a), and gave the Commission authority to issue rules and regulations necessary for protection of public health and safety, see 42 U.S.C. 2201, but Congress did not define "adequate protection", nor did it command the Commission to define it.

Such a definition would have to take one of two forms, one of them incapable of preventing the abuses the commenters are concerned about, and the other simply not possible yet. The first of these would be a verbal definition of the kind encountered in, for instance, the various "reasonable man" standards in the common law. After the pattern of these, the Commission could say, correctly, that "adequate protection" is not zero risk, that it is the same as "no undue risk", that it has long-term and short-term aspects, and that it is that level of safety which the Atomic Energy Act requires for initial and continued operation of a nuclear power plant. However, such a definition clearly will not, of itself, prevent the abuses UCS and NUBARG are concerned about, nor is such a standard sufficiently helpful to the NRC staff in actual practice.

Thus, if there is to be a useful and generally applicable definition of "adequate protection", it must take another, more precise form, namely, quantitative. Several of the commenters seem to have such a definition in mind when they call for "objective criteria", some "real numbers", and the like. In fact, the Commission is actively pursuing reliable quantitative measures of safety, and some quantitative and generally applicable definition of "adequate protection" may eventually

emerge as a byproduct of the Commission's efforts, still in their early stages, to implement its general safety goals, which take a partly quantitative form. (See 51 FR 30028; August 21, 1986, Policy Statement on Safety Goals.) However, given the state of the art in quantitative safety assessment, it is not reasonable to expect that the Commission could make licensing decisions—let alone decisions on whether to consider cost in backfitting— wholly on a quantitative definition of "adequate protection". Surprisingly, some of the commenters who call for "objective criteria", "some real numbers", and the like, have in the past criticized quantitative risk assessments.

Nonetheless, even in the absence of a useful and generally applicable definition of "adequate protection", the Commission can still make sound judgments about what "adequate protection" requires, by relying upon expert engineering and scientific judgment, acting in the light of all relevant and material information. As UCS itself said in its comments on the proposed 1985 revision of the rule, "(u)ltimately, the determination of what standards must be met in order to provide a reasonable assurance that the public health and safety will be protected comes down to the reasoned professional judgment of the responsible official."

The Commission's exercise of this judgment will take two familiar forms, of which the most important is rule and regulation. An essential point of the Commission's having regulations is to flesh out the "adequate protection" standard entrusted to the Commission by Congress. See UCS v. NRC, 824 F.2d at 117–18. Exercising engineering and scientific judgment in the light of all relevant and material information, the NRC identifies potential hazards and then requires that designs be able to cope with such hazards with sufficient safety margins and reliable backup systems. Regulations and guidance arrived at in this way do not, strictly speaking, "define" adequate protection, since there will be times when the NRC issues rules which require something beyond adequate protection. Nonetheless, compliance with such regulations and guidance may be presumed to assure adequate protection at a minimum. As the Commission has said on many occasions, compliance with the Commission's regulations and guidance "should provide a level of safety sufficient for adequate protection of the public health and safety and common defense and security under the Atomic Energy Act." (49 FR 47034, 47036,

col. 2, November 30, 1984, proposed 1985 rule; see also 50 FR 38097, 38101, col. 3, September 20, 1985, final 1985 rule; 51 FR 30028, col. 1, August 21, 1986, Policy Statement on Safety Goals.)

Because "adequate protection" is presumptively assured by compliance with the regulations and other license requirements, all the versions of the backfit rule—the 1970 rule, the 1985 rule, and the one set out in this document, see § 50.109(a)(4)(i)—have a "compliance" exception: plants out of compliance may be backfitted without findings of "substantial increase" in protection or a "justification" of costs.

However—and here is where the lack of a general definition for "adequate protection" poses a challenge— "adequate protection" is only presumptively assured by compliance. As the Commission said in promulgating the 1985 revision, the presumption may be overcome by, for instance, new information which indicates that improvements are needed to ensure adequate protection. (50 FR 38101, col. 3.) Such new information may reveal an unforeseen significant hazard or a substantially greater potential for a known one, or insufficient margins and backup capability. Engineering judgment may, in the light of such information, conclude that restoration of the level of protection presumed by the regulations requires more than compliance. Thus both the 1985 revision and the revision below contain exemptions for backfits necessary to assure "adequate protection", or, as the 1985 rule equivalently said, "no undue risk". See § 50.109(a)(4)(ii) of the rule set out in this document.

If compliance does not assure adequate protection, the Commission must be able to determine how much more protection is required, and a precise and generally applicable definition of "adequate protection" would facilitate that determination. But such a definition would have only a limited role to play. The first and most crucial question is whether the proposed backfit is required to bring a plant into compliance. Only if the proposed backfit requires more than compliance with NRC regulations and license conditions need there be a determination as to what "adequate protection" requires. Given this relation between compliance and "adequate protection", the industry might be more concerned than UCS is about the lack of a general definition of "adequate protection", for UCS will at least have the comfort of knowing that compliance will be secured before cost is considered, but the industry cannot be sure how much more than compliance may be asked of it despite the cost.

Where, as in the cases contemplated by the second exception provision of the rule, more than compliance is required and quantitative criteria do not define "adequate protection", the agency must fall back on the second familiar form in which engineering judgment is exercised by the Commission, namely, case-by-case. Administrative agencies are not required to proceed by rule alone, for the method of case-by-case judgment is quite capable of meeting the requirement that the factors on which administrative decisions are based be articulated. Rather than proceeding by an almost ministerial application of "objective criteria", the Commission must fashion a series of case-by-case judgments into a well-reasoned and factually well-supported body of decisions which, acting as reasoned precedent, can control and guide the Commission's exercise of the discretion granted it by Congress in precisely the way in which common-law precedents control and guide the common law judge's exercise of his or her judgment. See Nader v. Ray, 363 F.Supp. 946, 954–55 (D.D.C. 1973) (determining what constitutes adequate protection calls for exercise of discretion in a judgmental process very different from acting in accord with a clear, non-discretionary legal duty).

The Commission foresaw the need to proceed case-by-case on occasion and therefore made it a principal aim of the backfit rule to centralize the responsibility-and document the bases for case-by-case decisions for such decisions. The Commission thereby hoped to better assure that such decisions as might of necessity be case-by-case would form a reasoned and coherent body.[6]

---

[6] UCS alleges that in three instances the Commission has abused its discretion by applying cost considerations in specific cases where licensees are in compliance but adequate protection is at stake. However, UCS is misinformed about the first of the three cases, and its allegations about the other two reduce simple to disagreement over what constitutes adequate protection. We briefly discuss the three cases below.

Citing trade journal articles which quote unnamed NRC sources, UCS claims that the backfit rule caused the NRC staff to change its mind about requiring two licensees to conduct certain inspections and analyses in order to justify continued operations. The two plants in question had reactor pump coolant shafts similar to ones which elsewhere had shown a high probability of shearing off under certain conditions. UCS asserts that "[w]e * * * learn from this example the inherent lack of logic and circularity embedded in the rule: NRC is prevented, by operation of the rule, from asking questions needed to learn the degree of risk of a known equipment problem because they do not know the answers in advance."

Nothing in the Court's ruling in UCS v. NRC forbids the Commission's approach

However, the facts of the situations were not what UCS alleges them to have been; indeed the backfit rule was not involved. Letters were sent on April 23, 1986 requiring the licensees to submit within 20 days information which would "enable the Commission to determine whether or not [their] license(s) should be modified." Such information included information on design, operational history, schedules for inspection, plans for operator training, and "any analysis performed subsequent to those done for the FSAR (Final Safety Analysis Report) which would address the consequences of a locked rotor or broken shaft event during plant operation." These letters were sent under the first part of 10 CFR 50.54(f). This part authorizes such information requests without consideration of cost. As an earlier draft of the April 23 letter available in the NRC's Public Document Room shows, the NRC had planned to ask for new analyses under a later part of § 50.54(f) which authorizes requests not required to assure adequate protection if "the burden to be imposed * * * is justified in view of the potential safety significance of the issue to be addressed in the requested information." 10 CFR 50.54(f). (This "safety significance" standard, by its emphasis on "potential", requires less than is required by the "(actual) substantial increase" standard in the backfit rule and also avoids the circularity UCS alleges.) However, the staff sensibly opted for first asking whether such analyses had already been done. In fact they had, or were underway when the letters were sent. The backfit rule played no part here.

UCS' second instance of alleged abuse involves the Mark I containment, about whose performance in beyond-design-basis accidents (ones which involve damage to the reactor core) there is substantial uncertainty. UCS asserts that cost considerations have blocked staff action which would have brought about a significant reduction in some of the figures which estimate the probability that the Mark I would fail in certain kinds of beyond-design-basis accidents. UCS adds in passing that those figures represent undue risk. The NRC staff has already made a formal reply to similar charges of undue risk. See, e.g., Boston Edison Co. (Pilgrim Nuclear Generating Station). Interim Director's Decision under 10 CFR 2.206, DD–87–14, 26 NRC 87-106 (1987). Suffice it here to say that the NRC staff has by no means completed its considerations of the Mark I containment, but that, given present information, the staff has concluded that overall severe-accident risks at plants with Mark I containments are not undue. Id. at 104–106. UCS is content to put forward only unsupported assertions to the contrary. Thus the staff may legitimately consider cost when deciding whether to backfit the Mark I containments.

UCS' third allegation of abuse rehearses part of its February 10, 1987 § 2.206 Petition to the Commission for immediate action to relieve allegedly undue risks posed by nuclear power plants designed by the Babcock & Wilcox Company. The NRC's Director of Nuclear Reactor Regulation responded fully to the Petition, denying it, on October 19, 1987 (UCS' comments on the proposed backfit rule were submitted on October 13). See Director's Decision Under 10 CFR 2.206, DD–87–18, 26 NRC—(October 19, 1987). The Director concluded that "there are no substantial health and safety issues that would warrant the suspension or revocation of any license or permit for such facilities." Slip Opinion at 63. Simply because UCS disagrees with such conclusions does not mean that the Commission is misusing the "adequate protection" standard.

to "adequate protection". UCS boldly asserts that the proposed rule "completely fail[ed] to comport with the orders and directions of the Court of Appeals in UCS v. NRC", that the Court "could not have been more clear about the defects of the backfit rule", that the proposed revised rule "suffers from the exact same defects" as the one vacated, that, indeed, "the new proposal is even more devoid of objective guidance or criteria * * * than was its predecessor."

UCS' criticisms are based on part of a single paragraph in the Court's decision. In pertinent part, that paragraph says, "* * * In our view, the backfitting rule is an exemplar of ambiguity and vagueness; indeed, we suspect that the Commission designed the rule to achieve this very result. The rule does not explicate the scope or meaning of the three listed 'exceptions'. The rule does not explain the action the Commission will (in italics) take when a backfit falls within one of these exceptions. In short, the rule does not speak in terms that constrain the Commission from operating outside the bounds of the statutory scheme." 824 F.2d at 119.

UCS says that this portion of a paragraph was an "order" by the Court to get the Commission to "stop trying to obscure its intentions through ambiguous and vague language * * *." Whether the Court's language amounts to an "order" or only strong advice, we have followed it. For one thing, the rule explicitly says that backfits falling within the exceptions will be imposed (inexplicably, UCS asserts that the proposed rule did not have this provision). See § 50.109(a)(4). For another, both in what we have already said, and in what we shall be saying in response to NUBARG's comments on the exceptions provisions, we shall have explicated the scope and meaning of the three listed exceptions.

However, we have not taken the quoted language of the Court to mean that, after years of making rules and adjudicating cases which ultimately depend on the Commission's judgment about what "adequate protection" requires, the Commission should be obliged to give a mechanically applicable definition of "adequate protection" in order to avoid using the time-honored method of case-by-case, precedent-guided, judgment to implement only a part of the backfit rule. Certainly, the Court never even noted a lack of a general definition of "adequate protection" in the rule, let alone "ordered" the Commission to provide such a definition.

Appendix B   5

UCS' position lacks all sense of proportion. We must emphasize the core of the Court's decision, rather than get bogged down by transforming a suspicion and a few criticisms of the rule into an order to undertake an unprecedented task of definition.

Reviewing the exceptions in the rule, and various statements in the Federal Register notice accompanying the rule, the Court said, "We conceivably could read the terms of this rule to comply with the statutory scheme we have described above (that is, a scheme in which economic costs can play no part in establishing what adequate protection requires)." Id. Moreover, the Court says this despite the lack of any summary, general, "objective" definition of "adequate protection" in the rule.

But the Court then went on to say, "Statements that the Commission has made in promulgating the rule and in defending it before this court, however, disincline us from interpreting the rule in this fashion." Id. Again, it is not the lack of a definition of adequate protection that disinclined the Court from saving the rule, but rather certain statements the Commission had made which seemed to suggest that the Commission might consider economic cost when deciding what adequate protection required.

The Three Exceptions

Echoing the Court's remark that the rule "does not explicate the scope or meaning of the three listed 'exceptions'", id., NUBARG "believes that there is a substantial amount of overlap in these exceptions and that they have not been adequately defined or explained in the proposed rule." NUBARG and others representing the industry are concerned that the two exception provisions which use the phrase "adequate protection", §§ 50.109(a)(4) (ii) and (iii), may "swallow" the rule. One industry commenter objects to the notion, implied by § 50.109(a)(4)(ii), that adequate protection might require more than compliance. Another is concerned that § 50.109(a)(4)(iii), the exception which has been added in response to the Court's ruling, might lead to redefinitions of "adequate protection" that would threaten loss of licenses.

To avoid these results, NUBARG and others recommend deleting one of the two exception provisions which use the phrase "adequate protection". NUBARG's choice is § 50.109(a)(4)(ii), retained from the 1985 version of the rule, where it used the equivalent phrase, "no undue risk". This section provides that the "substantial increase"

and "costs justified" standards will not apply to backfits necessary to provide adequate protection to public health and safety. NUBARG calls this provision redundant to the exception for backfits required for the sake of compliance, § 50.109(a)(4)(i). As was noted above, NUBARG reports that its research has uncovered no case in which the Commission "has recognized that some additional measures not contained in existing requirements are necessary to ensure that a facility continues to meet the current level of adequacy." Two other commenters believe that the exception provision added because of the litigation, § 50.109(a)(4)(iii), should be deleted, as being redundant to the provision NUBARG would like to see deleted.

No matter which of the two provisions the commenter would like to see deleted, the commenter would like some restrictions placed on the use of the remaining one. The restriction by far the most frequently proposed is that no action may be taken under the remaining exception provision in the absence of "significant new information or the occurrence of an event which clearly shows" that the action is necessary.

In sum, these commenters either reopen an issue settled in 1985 or they recommend deleting that part of the rule which directly responds to the Court's ruling. We take neither course, for, even putting the 1985 rule and the Court's ruling aside, if either of the two provisions were to be deleted, an essential power of the Commission would be remain unimplemented.

First, the exception for backfits necessary to secure adequate protection, § 50.109(a)(4)(ii), must be retained, because it must be made clear that Commission action is not to be obstructed by cost considerations in a situation where compliance has indeed proved to be insufficient to secure the level of protection presumed in the rule, order, or commitment in question. Despite the results of NUBARG's research, such situations have arisen. See, e.g., SECY-88-346, "Evaluation of Managing Plant-Specific Backfit Requirements", November 21, 1988. Accordingly, this exception provision is not redundant to the exception for backfits necessary to restore compliance. Neither is it redundant to the exception for backfits involving the defining or redefining of "adequate protection", for the latter exception assumes some change in the NRC's judgment of what level of protection should be regarded as "adequate".

Retaining § 50.109(a)(4)(ii) will not give the Commission the power to proclaim at will that compliance is not enough. As we said in the statement of considerations accompanying the 1985 rule, and have in part reiterated in the response to UCS' comments, the regulations, though they do not define "adequate protection", are presumed to ensure it, and, in the absence of a redefinition of "adequate protection", that presumption can be overcome only by significant new information or some showing that the regulations do not address some significant safety issue. "(I)t may be presumed that the current body of NRC safety regulations provides adequate protection. Where new information indicates that improvements are needed to ensure there is 'no undue risk' on * * * a * * * basis which the Commission believes to be the minimum necessary, such requirements must be imposed." (50 FR at 38101–102.)

Second, the exception provision for backfits which are necessary under a defining or redefining of "adequate protection", § 50.109(a)(4)(iii), must be retained because it must be made clear that, as the Court held, cost may not be a factor in setting the level of protection judged as "adequate".[1] As NUBARG acknowledges, citing Power Reactor Development Co. v. International Union of Electrical, Radio and Machine Workers, AFL-CIO, 367 U.S. 396, 408 (1961), the Commission has both the power to define "adequate protection", and the power to re-define it.[2] Without this last exception provision, it might appear from the rule either that the Commission had no such power or that it was restricted by cost considerations, contrary to the Court's ruling. Nor should this exception provision be limited to situations involving "significant new information," as proposed in several comments.

This last exception may be thought by some to threaten to swallow the backfit rule. We believe, however, that instances of backfits based on a "redefinition" of "adequate protection" will be rare. Moreover, the case-by-case approach which is required in the

---

[1] As the rule notes in § 50.109(a)(7), cost may nonetheless be a consideration in choosing the means of achieving "adequate protection".

[2] The words "defining or redefining" in this third exception should not be construed necessarily to mean "providing a useful and generally applicable definition", at least not until such a definition becomes possible. Under present conditions, the Commission will have "defined or redefined what level of protection is to be regarded as adequate" if it makes a judgment that, although compliance assures the level of protection that had been thought of as adequate, that level of protection should no longer be considered adequate.

absence of a general definition of "adequate protection" provides licensees—and the public—a large measure of protection from arbitrary action by the Commission. Citing case law, NUBARG says that, in applying this last exception provision, the Commission "must act rationally and consistently in light of available evidence", and "must apply a reasoned analysis indicating the prior policies and standards are being changed, not casually ignored * * *." We wholly agree, and believe that the approach envisioned by the backfit rule will facilitate the Commission's acting accordingly.

Other Matters

Two other comments bearing on the phrase "adequate protection" require an explicit response. First, several commenters from the industry would prefer that the rule state that the "documented evaluation" which the NRC must prepare in connection with any action under one of the exception provisions, see § 50.109(a)(4), should include consideration of as many of the factors which § 50.109(c) requires of a "backfit analysis" as are appropriate.

The suggested modification of the rule would have only limited utility. Few of the factors listed in § 50.109(c) of the rule are appropriate for consideration in a documented evaluation justifying action under the compliance exception in the rule. It is true that several of the factors in § 50.109(c), indeed, all of them but those in paragraphs (c) (5) and (7) and some of those in paragraph (c)(8) are appropriate for consideration under the "adequate protection" exception, to the extent that they require a showing of exactly what the licensees must do and a showing that the backfit in question actually contributes to safety. However, the Commission believes that the rule's requirement that the documented evaluation "include a statement of the objectives of and reasons for the modification and the basis for invoking the exception" adequately assures that the factors in § 50.109(c) will be considered to the extent relevant, without their being listed and labeled as if they were a part of a § 50.109(c) analysis. Thus, little, if anything, is to be gained by an explicit requirement that § 50.109(c) factors be considered in a documented evaluation.

Second, one citizens' group asserts that the backfit rule should not apply to rulemaking. This issue was thoroughly discussed in 1985. However, this group's comment puts the issue in a slightly altered light, and provides another opportunity to clarify the meaning of "adequate protection". The group argues

that since rules "define" "adequate protection", the Commission cannot apply the rule's "substantial increase" and "cost justified" standards in rulemaking without applying cost considerations in setting the standard of adequate protection, contrary to the Court's holding.

The answer to this comment is, of course, that the rules do not, strictly speaking, "define" "adequate protection", and they only presumptively assure it. Not only may there, as stated above, be individual cases that require actions that go beyond what is necessary under the regulations to assure adequate protection, there will also be times when the NRC issues a rule which requires something beyond adequate protection. This follows directly from the Commission's power under section 161 of the Atomic Energy Act, affirmed by the Court, to issue rules or orders to "minimize danger to life or property." See 42 U.S.C. 2201; see also USC v. NRC, 824 F.2d at 118. If a proposed rule requires something more than adequate protection, applying a cost standard to the proposed rule will not be introducing cost considerations into the setting of the adequate protection standard and is therefore permitted. Of course if the rule is directed at either establishing what level of protection is "adequate" or assuring that such a level of protection is met, then cost will play no role.

The backfit rule as set out below is substantially the same as the rule proposed in the Federal Register. (See 52 FR 34223; September 10, 1987.) Provisions which appeared at the end of § 50.109(a)(4) of the proposed rule, or in the footnote to that paragraph, appear below in new paragraphs (a) (5) through (7).

Environmental Impact: Categorical Exclusion

The NRC has determined that this final rule is the type of action described in categorical exclusion 10 CFR 51.22(c)(3). Therefore, neither an environmental impact statement nor an environmental assessment has been prepared for this final rule.

Paperwork Reduction Act Statement

This final rule does not contain a new or amended information collection requirement subject to the Paperwork Reduction Act of 1980 (44 U.S.C. 3501 et seq.). Existing requirements were approved by the Office of Management and Budget, Approval Number 3140–0011.

Regulatory Analysis

The revision to 10 CFR 50.109 will bring it into conformance with the holding in Union of Concerned Scientists, et al., v. U.S. Nuclear [1] Regulatory Commission, D.C. Cir. Nos. 85–1757 and 86–1219 (August 4, 1987). The revision clarifies the backfit rule to reflect NRC practice that, in determining whether to adopt a backfit requirement, economic costs will be considered only when addressing those backfits involving safety requirements beyond those needed to ensure the adequate protection of public health and safety. Such costs are not considered when establishing the adequate protection of public health and safety. This revised rule does not have a significant impact on State and local governments and geographical regions, public health and safety, or the environment; nor does it represent substantial costs to licensees, the NRC, or other Federal agencies. This constitutes the regulatory analysis for this rule.

Regulatory Flexibility Act Certification

In accordance with the Regulatory Flexibility Act of 1980, 5 U.S.C. 605(b), the Commission hereby certifies that this final rule, if promulgated, will not have a significant economic impact on a substantial number of small entities. The affected facilities are licensed under the provisions of 10 CFR 50.21(b) and 10 CFR 50.22. The companies that own these facilities do not fall within the scope of "small entities" as set forth in the Regulatory Flexibility Act or the Small Business Size Standards set forth in regulations issued by the Small Business Administration in 13 CFR Part 121.

Backfit Analysis

The NRC has determined that a backfit analysis is not required for this rule because it does not impose requirements on 10 CFR Part 50 licensees.

List of Subjects in 10 CFR Part 50

Antitrust, Classified information, Fire prevention, Incorporation by reference, Intergovernmental relations, Nuclear power plants and reactors, Penalty, Radiation protection, Reactor siting criteria, Reporting and Recordkeeping requirements.

For the reasons set out in the preamble and under the authority of the Atomic Energy Act of 1954, as amended, the Energy Reorganization Act of 1974, as amended, and 5 U.S.C. 552 and 553, the NRC is adopting the following amendments to 10 CFR Part 50.

PART 50—DOMESTIC LICENSING OF PRODUCTION AND UTILIZATION FACILITIES

1. The authority citation for Part 50 is revised to read as follows:

Authority: Secs. 102, 103, 104, 105, 161, 182, 183, 186, 189, 68 Stat. 936, 937, 938, 948, 953, 954, 955, 956, as amended, sec. 234, 83 Stat. 1244, as amended (42 U.S.C. 2132, 2133, 2134, 2135, 2201, 2232, 2233, 2236, 2239, 2282); secs. 201, as amended, 202, 206, 88 Stat. 1242, as amended, 1244, 1246 (42 U.S.C. 5841, 5842, 5846).

Section 50.7 also issued under Pub. L. 95-601, sec. 10, 92, Stat. 2951 (42 U.S.C. 5851). Section 50.10 also issued under secs. 101, 185, 6J Stat. 936, 955, as amended (42 U.S.C. 2131, 2235); sec. 102, Pub. L. 91-190, 83 Stat. 853 (42 U.S.C. 4332). Sections 50.23, 50.35, 50.55, and 50.56 also issued under sec. 185, 68 Stat. 955 (42 U.S.C. 2235). Sections 50.33a, 50.55a and Appendix Q also issued under sec. 102, Pub. L. 91-190, 83 Stat. 853 (42 U.S.C. 4332). Sections 50.34 and 50.54 also issued under sec. 204, 88 Stat. 1245 (42 U.S.C. 5844). Sections 50.58, 50.91, and 50.92 also issued under Pub. L. 97-415, 96 Stat. 2073 (42 U.S.C. 2239). Section 50.78 also issued under sec. 122, 68 Stat. 939 (42 U.S.C. 2152). Section 50.80-50.81 also issued under sec. 184, 68 Stat. 954, as amended (42 U.S.C. 2234). Section 50.103 also issued under sec. 108, 68 Stat. 939, as amended (42 U.S.C. 2138).

Appendix F also issued under sec. 187, 68 Stat. 955 (42 U.S.C. 2237).

For the purposes of sec. 223, 68 Stat. 958, as amended (42 U.S.C. 2273); §§ 50.10 (a), (b), and (c), 50.44, 50.46, 50.48, 50.54, and 50.80(a) are issued under sec. 161b, 68 Stat. 948, as amended (42 U.S.C. 2201(b)); §§ 50.10 (b) and (c), and 50.54 are issued under sec. 161i, 68 Stat. 949, as amended (42 U.S.C. 2201(i)); and §§ 50.9, 50.55(e), 50.59(b), 50.70, 50.71, 50.72, 50.73, and 50.78 are issued under sec. 161o, 68 Stat. 950, as amended (42 U.S.C. 2201(o)).

2. Section 50.109 is revised to read as follows:

§ 50.109 Backfitting.

(a)(1) Backfitting is defined as the modification of or addition to systems, structures, components, or design of a facility; or the design approval or manufacturing license for a facility; or the procedures or organization required to design, construct or operate a facility; any of which may result from a new or amended provision in the Commission rules or the imposition of a regulatory staff position interpreting the Commission rules that is either new or different from a previously applicable staff position after:

(i) The date of issuance of the construction permit for the facility for facilities having construction permits issued after October 21, 1985; or

(ii) Six months before the date of docketing of the operating license application for the facility for facilities having construction permits issued before October 21, 1985; or

(iii) The date of issuance of the operating license for the facility for facilities having operating licenses; or

(iv) The date of issuance of the design approval under Appendix M, N, or O of this part.

(2) Except as provided in paragraph (a)(4) of this section, the Commission shall require a systematic and documented analysis pursuant to paragraph (c) of this section for backfits which it seeks to impose.

(3) Except as provided in paragraph (a)(4) of this section, the Commission shall require the backfitting of a facility only when it determines, based on the analysis described in paragraph (c) of this section, that there is a substantial increase in the overall protection of the public health and safety or the common defense and security to be derived from the backfit and that the direct and indirect costs of implementation for that facility are justified in view of this increased protection.

(4) The provisions of paragraphs (a)(2) and (a)(3) of this section are inapplicable and, therefore, backfit analysis is not required and the standards in paragraph (a)(3) of this section do not apply where the Commission or staff, as appropriate, finds and declares, with appropriated documented evaluation for its finding, either:

(i) That a modification is necessary to bring a facility into compliance with a license or the rules or orders of the Commission, or into conformance with written commitments by the licensee; or

(ii) That regulatory action is necessary to ensure that the facility provides adequate protection to the health and safety of the public and is in accord with the common defense and security; or

(iii) That the regulatory action involves defining or redefining what level of protection to the public health and safety or common defense and security should be regarded as adequate.

(5) The Commission shall always require the backfitting of a facility if it determines that such regulatory action is necessary to ensure that the facility provides adequate protection to the health and safety of the public and is in accord with the common defense and security.

(6) The documented evaluation required by paragraph (a)(4) of this section shall include a statement of the objectives of and reasons for the modification and the basis for invoking the exception. If immediately effective regulatory action is required, then the documented evaluation may follow rather than precede the regulatory action.

(7) If there are two or more ways to achieve compliance with a license or the rules or orders of the Commission, or with written licensee commitments, or there are two or more ways to reach a level of protection which is adequate, then ordinarily the applicant or licensee is free to choose the way which best suits its purposes. However, should it be necessary or appropriate for the Commission to prescribe a specific way to comply with its requirements or to achieve adequate protection, then cost may be a factor in selecting the way, provided that the objective of compliance or adequate protection is met.

(b) Paragraph (a)(3) of this section shall not apply to backfits imposed prior to October 21, 1985.

(c) In reaching the determination required by paragraph (a)(3) of this section, the Commission will consider how the backfit should be scheduled in light of other ongoing regulatory activities at the facility and, in addition, will consider information available concerning any of the following factors as may be appropriate and any other information relevant and material to the proposed backfit:

(1) Statement of the specific objectives that the proposed backfit is designed to achieve;

(2) General description of the activity that would be required by the licensee or applicant in order to complete the backfit;

(3) Potential change in the risk to the public from the accidental off-site release of radioactive material;

(4) Potential impact on radiological exposure of facility employees;

(5) Installation and continuing costs associated with the backfit, including the cost of facility downtime or the cost of construction delay;

(6) The potential safety impact of changes in plant or operational complexity, including the relationship to proposed and existing regulatory requirements;

(7) The estimated resource burden on the NRC associated with the proposed backfit and the availability of such resources;

(8) The potential impact of differences in facility type, design or age on the relevancy and practicality of the proposed backfit;

(9) Whether the proposed backfit is interim or final and, if interim, the justification for imposing the proposed backfit on an interim basis.

(d) No licensing action will be withheld during the pendency of backfit analyses required by the Commission's rules.

(e) The Executive Director for Operations shall be responsible for implementation of this section, and all analyses required by this section shall be approved by the Executive Director for Operations or his designee.

Dated at Rockville, Maryland, this 31st day of May, 1988.

For the Nuclear Regulatory Commission.

Samuel J. Chilk,

Secretary of the Commission.

[FR Doc. 88–12624 Filed 6–3–88; 8:45 am]

BILLING CODE 7590-01-M

DEPARTMENT OF THE TREASURY

Comptroller of the Currency

12 CFR Part 4

[Docket No. 88–8]

Description of Office, Procedures, Public Information; Deputy Chief Counsel (Operations) et al.

AGENCY: Comptroller of the Currency, Treasury.

ACTION: Final rule.

SUMMARY: The structure of the Law Department of the Office of the Comptroller of the Currency ("OCC") has recently been changed. This final rule sets forth the new descriptions for the positions of Deputy Chief Counsel (Operations) and Deputy Chief Counsel (Policy).

EFFECTIVE DATE: June 6, 1988.

FOR FURTHER INFORMATION CONTACT: Ferne Fisherman Rubin, Attorney, Legal Advisory Services Division, (202) 447–1880, Office of the Comptroller of the Currency, 490 L'Enfant Plaza East, SW., Washington, DC 20219.

SUPPLEMENTARY INFORMATION: On April 6, 1988, the OCC's Chief Counsel announced certain changes to the positions of Deputy Chief Counsel (Operations) and Deputy Chief Counsel (Policy); this amendment reflects these changes.

Notice and Comment

The OCC has determined that notice and comment are unnecessary under 5 U.S.C. 553(b)(3)(A) since this final rule pertains to rules of agency organization and procedure.

Reason for Immediate Effective Date

This final rule informs the public about a change in the Law Department's organization that has already occurred. Confusion could result if the proper position descriptions are not employed immediately.

Regulatory Flexibility Act

A Regulatory Flexibility Analysis is required only for rules issued for notice and comment. Because this final rule pertains to office organization and is therefore exempt from notice and comment procedures, no Regulatory Flexibility Analysis will be prepared.

Executive Order 12291

Section 1(a)(3) of Executive Order 12291 exempts from the requirements that a Regulatory Impact Analysis be prepared those regulations related to agency organization, management or personnel. Since this final rule is so classified, no Regulatory Impact Analysis is required.

List of Subjects in 12 CFR Part 4

National banks, Organization and functions (government agencies), Public information, Official forms, District offices, Field offices, Procedures, Delegation.

For the reasons given in the preamble, Part 4 of Chapter I, Title 12 of the Code of Federal Regulations is amended as follows:

PART 4—DESCRIPTION OF OFFICE, PROCEDURES, PUBLIC INFORMATION

1. The authority citation for Part 4 continues to read as follows:

Authority: 12 U.S.C. 1 *et seq.*, 5 U.S.C. 552, unless otherwise noted.

2. In Part 4, § 4.1a is amended by revising paragraph (a) (20) and (21) to read as follows:

§ 4.1a Central and field organization; delegations.

(a) * * *

(20) *Deputy Chief Counsel (Operations).* The Deputy Chief Counsel (Operations) is responsible for Law Department administration, the District Counsels, and the Legislative and Regulatory Analysis Division of the Law Department.

(21) *Deputy Chief Counsel (Policy).* The Deputy Chief Counsel (Policy) is responsible for the Enforcement and Compliance, Legal Advisory Services, Litigation, and Securities and Corporate Practices Divisions of the Law Department.

* * * * *

Date: May 27, 1988.

Robert L. Clarke,

Comptroller of the Currency.

[FR Doc. 88–12605 Filed 6–3–88; 8:45 am]

BILLING CODE 4810-33-M

FEDERAL HOME LOAN BANK BOARD

12 CFR Part 563

[No. 88–427]

Miscellaneous Conforming and Technical Amendments

Date: May 31, 1988.

AGENCY: Federal Home Loan Bank Board.

ACTION: Final rule; miscellaneous conforming and technical amendments.

SUMMARY: The Federal Home Loan Bank Board ("Board"), as the operating head of the Federal Savings and Loan Insurance Corporation ("FSLIC"), is amending its regulations in order to correct typographical and other technical errors, and to correct a reference to the Board's recordkeeping requirements with respect to accounts held in institutions the deposits of which are insured by the FSLIC ("insured institutions").

EFFECTIVE DATE: June 6, 1988.

FOR FURTHER INFORMATION CONTACT: Jerome L. Edelstein, (202) 377–7057, Deputy Director; or Carol J. Rosa, (202) 377–7037, Paralegal Specialist, Regulations and Legislation Division, Office of General Counsel, Federal Home Loan Bank Board, 1700 G Street NW., Washington, DC 20552.

SUPPLEMENTARY INFORMATION: On August 15, 1986, the Board adopted final amendments expanding and clarifying its regulation concerning basic loan records that institutions chartered by the Board or insured institutions and their service corporations are required to maintain. 51 FR 30848 (August 29, 1986). One of the amendments revised 12 CFR 563.17–1(c) by providing that records related to accounts held in insured institutions reflect the Board's recent deletion of the requirement that for insurance of accounts purposes the insured institution's records disclose the names of the settlor (grantor) and trustee of a trust and contain a signature card for the trust executed by the trustee. The Board's deletion of this recordkeeping requirement was adopted on April 4, 1986. 51 FR 12122 (April 9, 1986). The April 1986 revision of 12 CFR 564.2 to delete paragraph (b)(3) was intended to decrease the recordkeeping requirements associated with obtaining trust account insurance coverage and to expedite settlement of insurance claims on such accounts. This amendment was not intended to apply to loan recordkeeping requirements of an insured institution or its service corporations but only to insurance

APPENDIX C

CHARTER

COMMITTEE TO REVIEW GENERIC REQUIREMENTS

(Revision 4, April 1987)

TABLE OF CONTENTS

                                                                          Page

I.   Purpose..   1

II. Membership..   2

III. CRGR Scope..   2

IV. CRGR Operating Procedures...   5

V.   Reporting Requirements..   9

Attachment 1:   New Generic Requirement and Staff Position Review Process

Attachment 2:   Procedures to Control Communication of Generic Requirements and
                Staff Positions to Reactor Licensees

APPROVED BY THE COMMISSION JUNE 16, 1982 (SECY-82-39A)

REVISION 1 APPROVED BY THE COMMISSION (SECY MEMO DTD JANUARY 6, 1984)

REVISION 2 APPROVED BY THE COMMISSION (COMSECY-86-5, JUNE 20, 1986)

REVISION 3 APPROVED BY THE COMMISSION (SECY MEMO DTD AUGUST 13, 1986)

REVISION 4 APPROVED BY THE EDO (MEMO TO COMMISSIONERS, APRIL 6, 1987)

I.   <u>PURPOSE</u>

The Committee to Review Generic Requirements (CRGR) has the responsibility to
review and recommend to the Executive Director for Operations (EDO) approval or
disapproval of requirements or staff positions to be imposed by the NRC staff
on one or more classes of power reactors.   This review applies to staff proposals of requirements or positions which reduce existing requirements or positions and proposals which increase or change requirements.   The implementation
of this responsibility shall be conducted in such a manner so as to assure that
the provisions of 10 CFR 2.204, 10 CFR 50.109 and 10 CFR 50.54(f) as pertaining
to generic requirements and staff positions are implemented by the staff.   The
objectives of the CRGR process are to eliminate or remove any unnecessary burdens placed on licensees, reduce the exposure of workers to radiation in implementing some of these requirements, and conserve NRC resources while at the
same time assuring the adequate protection of the public health and safety and
furthering the review of new, cost-effective requirements and staff positions.
The CRGR and the associated staff procedures will assure NRC staff implementation of 10 CFR 50.54(f) and 50.109 for generic backfit matters.   The overall
process will assure that requirements and staff positions in place or to be
issued (a) do in fact contribute effectively and significantly to the health
and safety of the public, and (b) do lead to utilization of both NRC and
licensee resources in as optimal a fashion as possible in the overall achieve-
ment of protection of public health and safety.   By having the Committee submit
recommendations directly to the EDO, a single agencywide point of control will
be provided.

The CRGR will focus primarily on proposed new requirements and staff positions,
but it will also review selected existing requirements and staff positions
which may place unnecessary burdens on licensee or agency resources.   In reaching its recommendation, the CRGR shall consult with the proposing office to
ensure that the reasons for the proposed requirement or staff position are well
understood and that the provisions of 10 CFR 50.109, 50.54(f), and 10 CFR
2.204, if applicable, are appropriately addressed by the staff proposal.   The
CRGR shall submit to the EDO a statement of the reasons for its recommendations.   This statement shall provide a clear indication of the basis for the
recommendation and, when appropriate, relate this basis to the provisions of 10
CFR 50.109, 50.54(f), and 10 CFR 2.204.

Tools used by the CRGR for scrutiny are expected to include cost-benefit analysis and probabilistic risk assessment where data for its proper use are adequate.   Therefore, to the extent possible, written staff justifications should
make use of these evaluation techniques.   The use of cost-benefit analyses and
other tools should help to make it possible to determine which proposed requirements and staff positions have real safety significance, as distinguished
from those proposed requirements and staff positions which should be given a
lower priority or those which might be dropped entirely.   When such techniques
cannot be applied for lack of available, appropriate, or relevant data, other
methods will be used.

The EDO may authorize deviations from this Charter when the EDO, after consulting with the Chairman, finds that such action is in the public interest
and the deviation otherwise complies with applicable regulations including

10 CFR 2.204, 50.54(f) and 50.109. Such authorization shall be written and shall become a part of the record of CRGR actions. The rulemaking proposal presented to and considered by the CRGR, and ultimately, if presented to the Commission, should include any necessary exemption request with supporting reasons for the proposed exemption.

II. MEMBERSHIP

This Committee shall be chaired by the Office Director, AEOD, and it shall consist of, in addition to the CRGR Chairman, one individual each from NRR, NMSS, the Regions, and RES appointed by the Executive Director for Operations and one individual from OGC appointed by the EDO with the concurrence of the General Counsel. The regional individual shall be selected from one of the regional offices, and this assignment shall be considered developmental, with a new selection made by the appointing official after that official judges that sufficient experience has been gained by the incumbent regional representative. The CRGR Chairman shall assure that process controls for overall agency management of the generic backfit process are developed and maintained. These process controls shall include specific procedures, training, progress monitoring systems, and provisions for obtaining and evaluating both staff and industry views on the conduct of the backfit process. The CRGR Chairman is also responsible for assuring that each licensee is informed of the existence and structure of the NRC program described in th s Charter. The CRGR Chairman shall assure that substantive changes in the Charter are communicated to all licensees.

AEOD will provide staff support. The Committee may use several non-NRC persons as consultants in special technical areas.

New members will be appointed as the need arises. If a member cannot attend a meeting of the CRGR, the applicable Office Director may propose an alternative for the appointing official's approval. It is the responsibility of the alternate member to be fully versed on the agenda items before the Committee.

III. CRGR SCOPE

A. The CRGR shall consider all proposed new or amended generic requirements and staff positions to be imposed by the NRC staff on one or more classes of power reactors. These include:

  (i) All staff papers which propose the adoption of rules or policy statements affecting power reactors or modifying any other rule so as to affect requirements or staff positions applicable to reactor licensees, including information required of reactor licensees or applicants for reactor licenses or construction permits.

  (ii) All staff papers proposing new or revised rules of the type described in paragraph (i), including Advanced Notices.

(iii) All proposed new or revised regulatory guides; all proposed new or revised Standard Review Plan (SRP) sections; all proposed new or revised branch technical positions; all proposed generic letters; all multiplant orders, show cause orders, and 50.54(f) letters[1]; all bulletins and circulars; and USI NUREGs; and all new or revised Standard Technical Specifications.

All staff proposed generic information requests will be examined by the CRGR in accordance with 10 CFR 50.54(f). Except for information sought to verify licensee compliance with the current licensing basis for a facility, the staff must prepare the reason or reasons for each information request prior to issuance to ensure that the burden to be imposed on respondents is justified in view of the potential safety significance of the issue to be addressed in the requested information. CRGR examination of generic letters will include those letters proposed to be sent to construction permit holders. For those plants for which an operating license is not yet issued, an exception to staff analysis may be granted by the Office Director only if the staff seeks information of a type routinely sought as part of the standard procedures applicable to the review of applications. If a request seeks to gather information pursuant to development of a new staff position, then the exception does not apply and the reasons for the request must be prepared and approved prior to issuance of the request. When staff evaluations of the necessity for a request are required, the evaluation shall include at least the following elements:

(a) A problem statement that describes the need for the information in terms of potential safety benefit.

(b) The licensee actions required and the cost to develop a response to the information request.

(c) An anticipated schedule for NRC use of the information.

B. The CRGR shall consider all licenses, license amendments, approvals of Preliminary Design Approvals (PDAs) and Final Design Approvals (FDAs), minutes of conferences with owners groups, licensees or vendors, staff approvals of topical reports, information notices, and all other documents, letters or communications of a generic nature which are presented

---

1   It is expected that the offices will develop internal procedures to ensure that information requests are developed in accordance with 50.54(f)

to reflect or interpret NRC staff positions, unless such documents refer only to requirements or staff positions[2] previously applicable to the affected licensees and approved by the appropriate officials. The following are examples of approved staff positions not requiring CRGR review:

(i)   positions or interpretations which are contained in regulations, policy statements, regulatory guides, the Standard Review Plan, branch technical positions, generic letters, orders, topical approvals, PDAs, FDAs, licenses and license amendments which have been promulgated prior to November 12, 1981. Any document or communication of this type shall cite and accurately state the position as reflected in a previously promulgated regulation, order, Regulatory Guide, SRP, etc.

(ii)   positions after November 12, 1981 which have been approved through this established generic review process.

C.   For those rare instances where it is judged that an immediately effective action is needed to ensure that facilities pose no undue risk to the health and safety of the public (10 CFR 50.109(a)(4)(ii)), no prior review by the CRGR is necessary. However, the staff shall conduct a documented evaluation which includes a statement of the objectives of and reasons for the actions and the basis for invoking the exception. The analysis referenced in 50.109(a)(2) may be conducted either before or after the action is taken and shall be subject to CRGR review. This analysis shall document the safety significance and appropriateness of the action taken and consideration of how costs contribute to selecting the solution among various acceptable alternatives. The CRGR Chairman should be notified by the Office Director originating the action. These immediately effective requirements will be reported to the Committee for information and will be included in the report to the Commission.

D.   For each proposed requirement or staff position not requiring immediately effective action, the proposing office is to identify the requirement as either Category 1 or 2.

Category 1 requirements and staff positions are those which the proposing office rates as urgent to overcome a safety problem requiring immediate resolution or to comply with a legal requirement for immediate or near-term compliance. Category 1 items are expected to be infrequent and few in number, and they are to be reviewed or otherwise dealt with within 2-working days of receipt by the CRGR. If the appropriateness of designation as Category 1 is questioned by the CRGR Chairman, and if the question is not resolved within the 2 working-day limit, the proposed

---

2   It is expected that the offices shall develop internal procedures to ensure that the documents and communications referenced above will contain only previously approved requirements or staff positions.

requirement or staff position is to be forwarded by the CRGR Chairman to
the EDO for decision.

Category 2 requirements and staff positions are those which do not meet
the criteria for designation as Category 1. These are to be scrutinized
carefully by the CRGR on the basis of written justification, which must
be submitted by the proposing office along with the proposed requirement
or staff position.

Staff proposed generic modifications considered necessary to bring facilities into compliance with licenses or the rules or orders of the
Commission, or into conformance with written commitments by licensees,
will not require analyses of the type described in Section IV (B)(vii).
The proposed action shall be presented to the CRGR Chairman with a documented evaluation including a statement of the objectives of and reasons
for the proposed requirement or staff position and the basis for involving the exception under 10 CFR 50.109(a)(4)(i).

E. The CRGR Chairman shall compile and maintain a list of projected generic
requirements and staff positions based on input from the NRC offices.
The CRGR may receive early briefings from the offices on the proposed
new generic requirements or staff positions before the staff has
developed the requirements or positions and held discussions with the
ACRS.

F. The CRGR may be consulted on any issue deemed appropriate by the CRGR
Chairman.

IV. CRGR OPERATING PROCEDURES

A. Meeting Notices

Meetings will generally be held at regular intervals and will be scheduled well in advance. Meeting notices will generally be issued by the
CRGR Chairman 2 weeks in advance of each meeting, except for Category 1
items, with available background material on each item to be considered
by the Committee.

B. Contents of Packages Submitted to CRGR

The following requirements apply for proposals to reduce existing requirements or positions as well as proposals to increase requirements or
positions. Each package submitted to the CRGR for review shall include
fifteen (15) copies of the following information:

(i) The proposed generic requirement or staff position as it is proposed to be sent out to licensees.

---

3 The requirements of the backfit rule and the Commission guidance for re-
laxation of requirements and staff positions shall continue to apply.

(ii) Draft staff papers or other underlying staff documents supporting the requirements or staff positions. (A copy of all materials referenced in the document shall be made available upon request to the CRGR staff. Any committee member may request CRGR staff to obtain a copy of any referenced material for his or her use.)

(iii) Each proposed requirement or staff position shall contain the sponsoring office's position as to whether the proposal would increase requirements or staff positions, implement existing requirements or staff positions, or would relax or reduce existing requirements or staff positions.

(iv) The proposed method of implementation along with the concurrence (and any comments) of OGC on the method proposed.

(v) Regulatory analyses generally conforming to the directives and guidance of NUREG/BR-0058 and NUREG/CR-3568.

(vi) Identification of the category of reactor plants to which the generic requirement or staff position is to apply (that is, whether it is to apply to new plants only, new OLs only, OLs after a certain date, OLs before a certain date, all OLs, all plants under construction, all plants, all water reactors, all PWRs only, some vendor types, some vintage types such as BWR 6 and 4, jet pump and nonjet pump plants, etc.).

(vii) For each such category of reactor plants, an evaluation which demonstrates how the action should be prioritized and scheduled in light of other ongoing regulatory activities. The evaluation shall document for consideration information available concerning any of the following factors as may be appropriate and any other information relevant and material to the proposed action:

 (a) Statement of the specific objectives that the proposed action is designed to achieve;

 (b) General description of the activity that would be required by the licensee or applicant in order to complete the action;

 (c) Potential change in the risk to the public from the accidental offsite release of radioactive material;

 (d) Potential impact on radiological exposure of facility employees and other onsite workers.

 (e) Installation and continuing costs associated with the action, including the cost of facility downtime or the cost of construction delay;

(f) The potential safety impact of changes in plant or operational complexity, including the relationship to proposed and existing regulatory requirements and staff positions;

(g) The estimated resource burden on the NRC associated with the proposed action and the availability of such resources;

(h) The potential impact of differences in facility type, design or age on the relevancy and practicality of the proposed action;

(i) Whether the proposed action is interim or final, and if interim, the justification for imposing the proposed action on an interim basis.

(viii) For each evaluation conducted pursuant to 10 CFR 50.109, the proposing office director's determination, together with the rationale for the determination based on the considerations of paragraphs (i) through (vii) above, that

(a) there is a substantial increase in the overall protection of public health and safety or the common defense and security to be derived from the proposal; and

(b) the direct and indirect costs of implementation, for the facilities affected, are justified in view of this increased protection.

(ix) For each evaluation conducted for proposed relaxations or decreases in current requirements or staff positions, the proposing office director's determination, together with the rationale for the determination based on the considerations of paragraphs (i) through (vii) above, that

(a) the public health and safety and the common defense and security would be adequately protected if the proposed reduction in requirements or positions were implemented, and

(b) the cost savings attributed to the action would be substantial enough to justify taking the action.

C.   CRGR Staff Review

CRGR staff shall review each package for completeness. If the package is not sufficient for CRGR consideration, it shall be returned by the CRGR Chairman to the originating office with reasons for such action. Prior notice to the Committee is not needed; however, CRGR members shall be informed of such actions.

- An accepted package shall be scheduled for CRGR consideration; however, scheduling priorities shall be at the discretion of the CRGR Chairman.

- All requests for particular scheduling shall be made to the CRGR Chairman.

- The CRGR staff may obtain additional information from industry and consultants on such proposals, particularly with respect to the cost of implementation, realistic schedule for implementation and the ability of licensees to safely and efficiently carry out the full range of safety-related activities at each facility while implementing the proposed requirement or staff position. The CRGR staff normally shall provide a brief summary analysis of each package to CRGR members prior to the meetings.

D. CRGR Meeting Minutes

At each meeting, for each package scheduled for discussion, the sponsoring office shall present to the CRGR the proposed generic requirement or staff position and respond to comments and questions. A reasonable amount of time, within the discretion of the CRGR Chairman, shall be permitted for discussion of each item by Committee members. At the conclusion of the discussion, each Committee member shall summarize his position. The minutes of each meeting, including CRGR recommendations and the bases therefor shall be prepared. Minutes normally shall be circulated to all members within 5-working days after the the meeting, and each member shall have 5-working days to comment in writing on the minutes. It s the responsibility of each member to assure that the minutes accurately reflect his views. All comments rece ved within that period shall be part of the minutes of the meeting.

The Committee shall recommend to the EDO, approval, disapproval, modification, or conditioning of generic proposals considered by the Committee, as well as the method of implementation of such requirements or staff positions and appropriate scheduling for such implementation, which shall give consideration to the ability of licensees to safely and efficiently carry out the entire range of safety-related activities at each facility. The minutes shall give an accurate description of the basis for the recommendations and shall accurately reflect the consensus decision of the Committee. Copies of the minutes shall be distributed to the Commission, Office Directors, Regional Administrators, CRGR Members, and the Public Document Room. The EDO's action taken in response to the Committee's recommendations shall be provided in writing to the Commission.

E. Recordkeeping System

The AEOD Assistant for CRGR Issues will assure that there is an archival system for keeping records of all packages submitted to the CRGR Chairman, actions by the staff, summary minutes of CRGR consideration of each package including corrections, recommendations by the Committee, and decisions by the EDO.

V. REPORTING REQUIREMENTS

The AEOD Assistant for CRGR Issues shall prepare a report to be submitted by the EDO to the Commission each month. The report will provide a brief summary of CRGR activities, including a list of all items that have been sent to the CRGR and their current status. The report shall be distributed to CRGR Members, Office Directors, Regional Administrators and the Public Document Room.

NEW GENERIC REQUIREMENT AND STAFF POSITION REVIEW PROCESS

The attached chart is a schematic representation of how new generic requirements and staff positions are developed, revised and implemented.

In the early stages of developing a proposed new requirement or staff position,
it is contemplated that the staff may have discussions with the industry, ACRS
and the public to obtain preliminary information of the costs and safety
benefits of the proposed action. On the basis of this information, the proposing office will prepare the package for CRGR review.

The CRGR may recommend approval, revision, or disapproval or that further
public comment be sought. After CRGR and EDO approval, there may be further
review by the ACRS or the Commission. Decisions by the Commission are
controlling.

Once final approval is received, the individual project managers will normally
work with each licensee to develop a plant-specific implementation schedule
taking into consideration all of the other requirements and staff positions that
are being implemented at .ach plant.

SCHEMATIC REPRESENTATION OF NEW REQUIREMENTS REVIEW

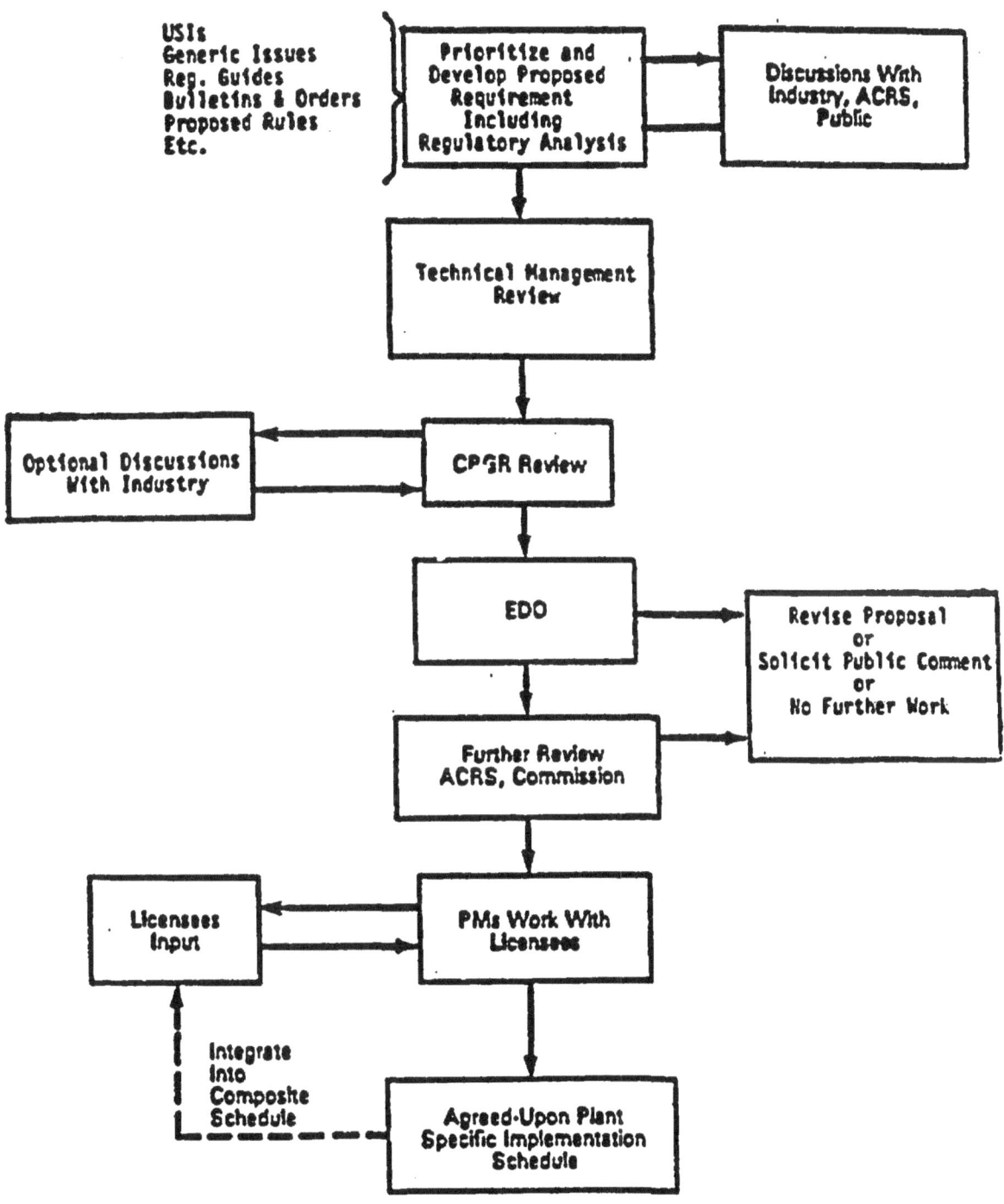

PROCEDURES TO CONTROL
GENERIC REQUIREMENTS AND STAFF POSITIONS

A. Background

In a memorandum from the Chairman to the Executive Director for Operations dated October 8, 1981, the Commission expressed concern over conflicting or inconsistent directives and requests to reactor licensees from various components of the NRC staff. By that memorandum, the Commission outlined certain recommended actions to establish control over the number and nature of requirements placed by NRC on reactor licensees. These included: establishing a Committee to Review Generic Requirements (CRGR); establishing a new position of Deputy Executive Director for Regional Operations and Generic Requirements (DEDROGR); conducting a survey of formal and informal mechanisms to communicate with reactor licensees; and developing and implementing procedures for controlling communications involving significant requirements covering one or more classes of reactors. In February 1987 the Commission approved a NRC reorganization that, among other changes, placed the CRGR operations under the Office of Analysis and Evaluation of Operational Data (AEOD). CRGR responsibilities and authorities were not directed to change under the new organizational structure; only the organizational location was changed. The following procedures have been established for controlling generic requirements or staff positions and are designed to implement the provisions of 10 CFR 50.109, 50.54(f) and 2.204.

B. Committee to Review Generic Requirements (CRGR)

Except for immediately effective actions, the CRGR shall review all proposed new generic requirements and staff positions to be imposed on one or more classes of power reactors in accordance with the Charter of the Committee, before such proposed requirements or staff positions are forwarded to the EDO and Commission and imposed on, or communicated for use or guidance to, any reactor licensee.

C. Office Responsibility

Each office shall develop internal procedures to assure that the following policy requirements regarding reactor licensees are carried out:

(1) All proposed generic requirements and staff positions (Table 1 attached) shall be submitted for CRGR review. Such submittals shall conform to the provisions of the CRGR Charter relating to the contents of such submittals.

(2) All generic documents, letters and communications that establish, reflect or interpret NRC staff positions or requirements (Table II attached) shall be submitted for review by CRGR unless these documents refer only to requirements or staff positions approved prior to November 12, 1981. In the latter case, the previously approved requirement or staff position should be specifically cited and accurately stated. Offices should be careful to

review new or specific interpretations to assure that they are only case-specific applications of existing requirements rather than initial applications having potential generic use. Case-specific applications are governed by NRC Manual Chapter 0514.

(3) For all other communications with licensees (Table III, attached), no statements shall be used that might suggest new or revised generic requirements, staff positions, guidance or recommendations unless such statements have been approved by the EDO or the Commission.

(4) In developing a proposed new generic requirement or staff position for CRGR review, an office may determine that it is in possession of important safety information that should be made available to licensees. It is the responsibility of that office to take immediate action to assure that such information is communicated to the licensees by the appropriate office. Such actions may be taken before completion of any proposed or ongoing CRGR reviews.

D. <u>Immediately Effective Action</u>

For those rare instances where it is judged that an immediately effective action is needed to ensure that facilities pose no undue risk to the health and safety of the public (10 CFR 50.109(a)(4)(ii)), no prior review by the CRGR is necessary. However, the staff shall conduct a documented evaluation which includes a statement of the objectives of and reasons for the actions and the basis for invoking the exception. The analysis referenced in 50.109(a)(2) may be conducted either before or after the action is taken and shall be subject to CRGR review. This analysis shall document the safety significance and appropriateness of the action taken and consideration of how costs contribute to selecting the solution among various acceptable alternatives. The CRGR Chairman should be notified by the Office Director originating the action. These immediately effective requirements will be reported to the Committee for information and will be included in the report to the Commission.

TABLE I

PRINCIPAL MECHANISMS USED BY NRC STAFF TO
ESTABLISH OR COMMUNICATE GENERIC REQUIREMENTS AND STAFF POSITIONS

Rulemaking[1]

    Advanced Notices
    Proposed Notices
    Final Rules
    Policy Statements

Other Formal Requirements[2]

    Multiplant orders including show cause orders and
     confirmatory orders

Staff Positions[3]

    Bulletins
    Circulars
    Multiplant letters (including 10 CFR 50.54f and TMI Action
     Plan letters)
    Regulatory Guides
    SRP (including Branch Technical Positions)
    Standard Tech Specs
    USI NUREGs

---

1   While Rulemaking is an action of the Commission rather than the staff, most rules are proposed or prepared by the staff.

2   The document itself imposes a legal requirement; e.g., regulatory orders or license conditions.

3   Documents that reflect staff positions which, unless complied with or a satisfactory alternative offered, the staff would impose or seek to have imposed by formal requirement.

TABLE II

MECHANISMS OFTEN USED TO INTERPRET GENERIC REQUIREMENTS OR STAFF POSITIONS

Action and Petitions for Rulemaking
--

Action on 10 CFR 2.206 Requests

Approval of Topicals

Facility Licenses and Amendments

SERs

FDAs, PDAs

I&E Manual

I&E (HQ) Positions

NUREG Reports (other than USIs)

Operator Licenses and Amendments

Single Plant Orders

Staff Positions on Code Committees

Unresolved Issues Resulting from Inspections

TABLE III

ADDITIONAL MECHANISMS SOMETIMES USED TO COMMUNICATE
GENERIC REQUIREMENTS OR STAFF POSITIONS

DES & FES

Entry, Exit and Management Meetings

Information Notices

Licensee Event Reports; Construction Deficiency Reports (Sent to Other Licensees)

NRC Operator Licensing People Contact with Licensees

Phone Calls or Site Visits by NRC Staff or Commission to Obtain Information (i.e., Corrective Actions, Schedules, Conduct Surveys, etc.)

Pleadings

Preliminary Notifications

Press Releases

Proposed Findings

Public Meetings, Workshops, Technical Discussions

Resident Inspector Day-to-Day Contact

SALP Reports

SECY Papers (Some Utilities Apparently Sent Operators to College Based on Recent SECY Paper on Operator Qualifications)

Special Reports

Speeches to Local Groups or Industry Associations

Technical Specifications

Telephone Calls and Meetings with Licensees, Vendors, Industry Representatives, Owners Groups

Testimony

APPENDIX D

U.S. NUCLEAR REGULATORY COMMISSION

MANUAL CHAPTER 0514

Volume: 0000   General Administration
Part   :   0500   Health and Safety   AEOD

CHAPTER 0514   NRC PROGRAM FOR MANAGEMENT OF PLANT-SPECIFIC
BACKFITTING OF NUCLEAR POWER PLANTS

0514-01   COVERAGE

011   This chapter establishes the requirements and guidance for NRC staff implementation of 10 CFR 50.109 and the provisions of 10 CFR 50 Appendix 0, 10 CFR 50.54(f), and 10 CFR 2.204, relating to plant-specific backfitting. Staff requirements and guidance for implementing the provisions of 10 CFR 50.109 pertaining to rules and other generic backfitting are beyond the scope of this chapter. Pertinent requirements and guidance for generic backfitting are contained in the CRGR Charter. Test and research reactor licensees are not covered by the provisions of the chapter.

012   This chapter defines the objectives, authorities, and responsibilities and establishes basic requirements for actions to be taken in instances where the NRC staff imposes new plant-specific regulatory staff positions on a nuclear power plant licensee.[1] This practice is commonly referred to as "backfitting" and for the purposes of this chapter is defined as the modification of or addition to systems, structures, components, or design of a facility; or the procedures or organization required to design, construct or operate a facility: any of which may result from a new or amended provision in the Commission rules or the imposition of a regulatory staff position interpreting the Commission rules that is either new or different from a previously applicable staff position. It should be clearly understood that backfits are expected to occur and are a part of the regulatory process to assure and improve the safety of nuclear power plants. However, it is important for sound and effective regulation that backfitting be conducted in a controlled process. Plant-specific backfitting is different from generic backfitting in that the former involves the imposition on a licensee of positions unique to a particular plant, whereas generic backfitting involves the imposition of the same or similar positions on two or more plants. This chapter governs those plant-specific backfits communicated to the licensees or identified by the licensees after July 6, 1988.

013   The management of plant-specific backfitting, for which guidance is provided in this document, does not relieve licensees from achieving and

---

[1]See Section 05 of this chapter for a definition of "licensee."

Approved: August 26, 1988

maintaining adequate protection of the public health and safety[2] or complying with the Commission's regulations, orders, license, or written licensee commitment. The management process is intended to provide disciplined NRC review of new or changed positions prior to imposing them.

The plant-specific backfit management process will enhance regulatory stability by assuring that changes in regulatory staff positions are in fact required to ensure that the facility provides adequate protection to public health and safety or to provide a substantial increase in the overall protection of the public health and safety or common defense and security. Such plant-specific backfitting is entirely proper given the agency's responsibility to ensure an adequate level of protection and the agency's authority to improve safety beyond this level.

0514-02 OBJECTIVES

021 It is the overall objective of this program to assure that plant-specific backfitting of nuclear power plants is justified and documented and to specify that the Executive Director for Operations is responsible for the proper implementation of the backfit process.

022 The specific objectives of this program are (a) to ensure that facilities provide adequate protection of the public health and safety; and (b) to allow for substantial improvements in the levels of protection of public health and safety beyond adequacy while avoiding any unwarranted burdens on the NRC, public or licensees in implementing backfits.

023 The program should assure to the extent possible that backfits to be issued will in fact contribute effectively and significantly to the health and safety of the public or the common defense and security. This objective is attained by assuring that plant-specific backfits will be communicated to the licensee only if necessary to provide an adequate level of safety, or after required regulatory analyses are completed and approved as described in Section 0514-042 of this chapter. The backfit and supporting regulatory analyses are approved by the appropriate Office Director or Deputy Director, or Regional Administrator or Deputy Regional Administrator, and forwarded to the Executive Director for Operations before the backfit and appropriate supporting analysis are communicated to the licensee.

0514-03 RESPONSIBILITIES AND AUTHORITIES

031 The Executive Director for Operations (EDO) is responsible to the Commission for plant-specific backfit actions. The EDO may review and modify any plant-specific backfit decision at his or her initiative or at the request

---

[2] Adequate protection of the public health and safety means the same as no undue risk and reasonable assurance of not endangering public health and safety. In NRC practice these standards are interchangeable.

Approved: August 26, 1988

of a licensee in accordance with Section 044. The EDO may authorize deviations from this chapter when the EDO finds that such action is in the public interest and the deviation otherwise complies with the applicable regulations.

032   The Director, Office for Analysis and Evaluation of Operational Data (AEOD), shall assure that process controls for overall agency management and oversight of the plant-specific backfit process are developed and maintained and shall coordinate the implementation of procedures within the other Offices and Regions. These process controls shall include specific procedures, training, progress monitoring systems, and provisions for obtaining and evaluating both staff and industry views on the conduct of the backfit process. The Director, AEOD, is also responsible for assuring that each licensee is informed of the existence and structure of the NRC program described in this chapter. The Director, AEOD, shall assure that substantive changes in the chapter and related procedures are communicated to the licensees.

033   The Director, Office of Nuclear Reactor Regulation (NRR), shall assure that an overall procedure for managing plant-specific backfitting that involves positions taken by NRR is developed, implemented, and maintained, in accordance with the chapter. The overall procedure shall be coordinated with AEOD and approved by the EDO. The Director, NRR, shall consult and coordinate with Regional Administrators and the Office of Nuclear Material Safety and Safeguards, as appropriate, to develop resolutions of proposed plant-specific backfits in program areas for which NRR has responsibility.

For backfits within NRR's program area of responsibility which are proposed by NRR staff, the Director or Deputy Director, NRR, without further delegation, shall approve the regulatory analysis prior to communicating the backfit and analysis to the licensee. For all backfits within the NRR program area of responsibility which are appealed by a licensee, the Director, NRR, shall make the decision on imposition of the backfit. The decision is subject to EDO review under Section 0514-031. The Director, NRR, shall assure NRR staff performance in accordance with this chapter.

034   The Director, Office of Nuclear Material Safety and Safeguards (NMSS), shall assure that an overall procedure for managing plant-specific backfitting that involves positions taken by NMSS is developed, implemented, and maintained, in accordance with this chapter. The overall procedure shall be coordinated with AEOD and approved by the EDO. The Director, NMSS, shall consult and coordinate with Regional Administrators and the Director of the Office of Nuclear Reactor Regulation, as appropriate, to develop resolutions of proposed plant-specific backfits in program areas for which NMSS activities may affect reactor plant licensees.

For backfits within the NMSS program area of responsibility which are proposed by NMSS staff, the Director or Deputy Director, NMSS, without further delegation, shall approve the regulatory analysis prior to communicating the backfit and analysis to the licensee. For all backfits within the NMSS program area of responsibility which are appealed by a licensee, the Director, NMSS, shall make the decision on imposition of the backfit. The decision is subject to EDO review under Section 0514-031. The Director, NMSS, shall assure NMSS staff performance in accordance with this chapter.

Approved: August 26, 1988

035 <u>Regional Administrators</u> shall assure that an overall procedure for managing plant-specific backfitting that involves positions taken by a Region in any program area for which the Region has been delegated authority, is developed, implemented, and maintained, in accordance with the chapter. The overall procedure shall be coordinated with AEOD and approved by the EDO.

Regional Administrators shall consult and coordinate with the Directors of the Offices of Nuclear Reactor Regulation and Nuclear Material Safety and Safeguards, as appropriate, to identify issues and develop resolutions of proposed plant-specific backfits where such backfitting would result from positions taken by the Region.

For backfits proposed by the Region, the Regional Administrator or Deputy Regional Administrator, without further delegation, shall approve the regulatory analysis prior to communicating the backfit and analysis to the licensee. For backfits proposed by the Region and appealed by the licensee, the Administrator is responsible for the conduct of the appeal process within the Region; however, if agreement cannot be reached at the Regional level, the decision on imposition of the backfit shall be made by the Director of the program office having responsibility for the program area relevant to the backfit. The decision is subject to EDO review under Section 0514-031. Each Regional Administrator shall assure Regional staff performance in accordance with this chapter.

036 <u>The Directors, Offices of Nuclear Reactor Regulation, and Nuclear Material Safety and Safeguards, and Regional Administrators,</u> shall approve regulatory analyses initiated by their staff members, who propose backfits within other program office areas of responsibility which have been delegated to them for implementation and decision authority, prior to communicating the backfit and analysis to the licensee.

037 <u>The Director, Office of Administration and Resources Management,</u> shall, in coordination with the Office Directors, and Regional Administrators, develop and maintain the overall NRC data base management system identified and described in Section 046 of this chapter.

038 NRC staff positions may be identified as potential backfits either by NRC staff or by persons who are not members of the NRC staff. Such identifications will be considered by the Office Director/Regional Administrator having responsibility to develop staff positions on the matter at issue. This Office Director/Regional Administrator will be responsible to make the determination as to whether the staff position is a backfit and whether the proposed backfit should be imposed on the licensee.

0514-04 BASIC REQUIREMENTS

041 <u>Information Requests Pursuant to 10 CFR 50.54(f)</u>. Paragraph 10 CFR 50.54(f) authorizes the NRC to require its licensees to provide additional safety information to enable the Commission to determine whether or not a license should be modified, suspended, or revoked. This paragraph (as amended in 50 <u>FR</u> 38097) requires the NRC to justify such information requests by a

Approved: August 26, 1988

supporting analysis which finds that the burden to be imposed is justified in view of the potential safety significance of the issue to be addressed in the requested information. The exceptions to this requirement are as follows:

a. No finding is required whenever there is reason to believe that the public health and safety may not be adequately protected and safety information is needed to decide if this is the case and to take any necessary corrective action.

b. Concerning the review of applications for licenses or amendments, or the conduct of inspection activities, for plants under construction, no finding will be necessary if the staff seeks information of a type routinely sought as a part of the standard procedures concerning the review of applications. If the request is not part of routine licensing review (for example, if it seeks to gather information pursuant to development of a new staff position), a staff analysis of the reasons for the request and a finding must be prepared and approved prior to issuance.

c. Concerning licensing review or inspection activities for operating plants, information requests seeking to verify licensee compliance with the current licensing basis for the facility are exempt from the necessity to prepare the reason or reasons for the request and to make a finding. Requests for information to determine compliance with existing facility requirements including fact-finding reviews, inspections and investigations of accidents or incidents, usually are not made pursuant to Section 50.54(f), nor are such requests normally considered within the scope of the backfit rule or this chapter.

The Directors of NRR and NMSS and Regional Administrators shall develop internal office procedures to ensure that there is a rational basis for all information requests not clearly excepted from the finding, whether or not it is clear that backfit action would result from staff evaluation of the information supplied by the licensee. The request must be evaluated to determine whether the burden imposed by the information request is justified in view of the potential safety significance of the issue to be addressed. The information request and the staff evaluation must be approved by the cognizant Office Director or Regional Administrator prior to transmittal of the request for information to a licensee.

NRC staff evaluations of the necessity for an information request shall include at least the following elements:

a. A problem statement that describes the need for the information in terms of potential safety benefit.

b. The licensee actions required and the cost to develop a response to the information request.

c. An anticipated schedule for NRC use of the information.

Approved: August 26, 1988

042 Identifying Plant-Specific Backfits. The NRC staff shall be responsible for identifying proposed plant-specific backfits as defined by Section 05 of the chapter. The staff at all levels will evaluate any proposed plant-specific position with respect to whether or not the position qualifies as a proposed backfit pursuant to Section 05 of this chapter. No staff position should be communicated to a licensee unless the NRC official communicating that position has ascertained whether or not the position is to be identified as a backfit. NRC Appendix 0514 provides information to help in identifying backfits arising from selected staff activities. When a staff proposed position is identified as a backfit the staff should determine expeditiously whether the backfit is needed to ensure adequate protection of the public health and safety or to comply with Commission rules or orders, the license, or written licensee commitments. If, and only if, the backfit does not meet this test, the appropriate staff office should proceed promptly with the preparation of a regulatory analysis (Section 043) for approval in accordance with this chapter.

Economic cost can never be a consideration either in defining what is an adequate level of protection or in ensuring that an adequate level of protection is achieved and maintained.

The staff may, at any point in the development of the regulatory analysis, decide that further analysis is likely to show either that the proposed safety benefit is not likely to be substantial additional overall protection, or that the direct and indirect costs of implementation are not likely to be justified. In this case, the issue may be closed, with appropriate notice sent to all parties and recorded in the recordkeeping system described in Section 046.

When (a) a staff proposed position is necessary to bring a facility into compliance with a license or the rules or orders of the Commission (Sections 052-a, 053-a), or into conformance with written commitments by the licensee (Sections 052-a, 053-b), or (b) the Director of NRR or NMSS determines that imposition of a backfit is necessary to ensure that the facility provides adequate protection to public health and safety, no regulatory analysis is required. Instead, the appropriate Director/Regional Administrator is to provide a documented evaluation to support the action taken.

The evaluation shall include a statement of the objectives of the reasons for the modification and the basis for invoking the exception. In the case of a backfit needed to assure that the facility provides adequate protection, the documented evaluation shall also include an analysis to document the safety significance and appropriateness of the action taken. Should it be necessary or appropriate for the Commission to prescribe a way to achieve adequate protection, the evaluation can include a consideration of how costs contribute to selecting the solution among various acceptable alternatives. However, cost will not be a factor in determining what constitutes an adequate level of protection. Such an evaluation is to be issued with the backfit except that, when an immediately effective regulatory action is necessary, and the safety need is so urgent that full documentation cannot be completed, the documentation may follow the backfit.

A proposed staff position which is not identified by the NRC staff as a backfit position may be claimed to be a backfit position by a licensee. The

Approved: August 26, 1988

staff will promptly consider a licensee claim of backfit to determine if the claimed backfit qualifies as such in accordance with Section 05 of this chapter. Licensees identifying such items should send a written claim of backfit (with appropriate supporting rationale) to the Office Director or Regional Administrator of the NRC staff person who issued the position with a copy to the EDO. If the NRC staff determination is that the issue is a backfit, the appropriate staff office should proceed immediately with the preparation of any required regulatory analysis for approval in accordance with this chapter.

If the determination is that the proposed staff position is not a backfit, the appropriate staff office shall document the basis for the decision and transmit it together with any documented evaluation required by this section to the licensee. In any case, the appropriate Office Director/Regional Administrator shall report to the EDO and inform the licensee, within 3 weeks after receipt of the written backfit claim, of the results of the determination and the plan for resolving the issue.

When a licensee is informed that a claimed backfit is, in the judgment of the NRC, not a backfit, the licensee may appeal this determination as described in Section 044 of this chapter.

043   Regulatory Analysis. Positions identified as plant-specific backfits requiring the regulatory analysis in this section shall be transmitted to licensees only after a determination that there is a substantial increase in the overall protection of the public health and safety or the common defense and security to be derived from the backfit, and that the direct and indirect costs of implementation for that facility are justified in view of the increased protection. The proposed backfit and supporting regulatory analysis must be approved by the appropriate Program Office Director or Deputy Director, or Regional Administrator or Deputy Regional Administrator and forwarded to the EDO before the backfit and its supporting regulatory analysis are transmitted to the licensee.

The regulatory analysis shall generally conform to the directives and guidance of NUREG/BR-0058 and NUREG/CR-3568, which are the NRC's governing documents concerning the need for preparation of regulatory analyses. In preparing regulatory analyses under this section, the staff should note that the complexity and comprehensiveness of an analysis should be limited to that necessary to provide an adequate basis for decisionmaking among the alternatives available. The emphasis should be on simplicity, flexibility, and common sense, both in terms of the type of information supplied and in the level of detail provided. The following information and any other information relevant and material to the backfit shall be included in the regulatory analysis, as available and appropriate to the analysis:

a.   A statement of the specific objective that the proposed backfit is designed to achieve. This should also include a succinct description of the backfit proposed, and how it provides a substantial increase in overall protection.

b.   A general description of the activity that would be required by the licensee in order to complete the backfit.

Approved: August 26, 1988

c. The potential safety impact of changes in plant design or operational complexity, including the relationship to proposed and existing regulatory requirements.

d. Whether the proposed backfit is interim or final and, if interim, the justification for imposing the proposed backfit on an interim basis.

e. A statement that describes the benefits to be achieved and the cost to be incurred. Information should be used to the extent that it is reasonably available, and a qualitative assessment of benefits may be made in lieu of the quantitative analysis where it would provide more meaningful insights, or is the only analysis practicable. This statement should include consideration of at least the following factors:

   (1) The potential change in risk to the public from the accidental offsite release of radioactive material.

   (2) The potential impact on radiological exposure of facility employees. Also consider the effects on other onsite workers, due both to installation of procedural or hardware changes and to the effects of the changes, for the remaining lifetime of the plant.

   (3) The installation and continuing costs associated with the backfit, including the cost of facility downtime or the cost of construction delay.

   (4) The estimated resource burden on the NRC associated with the proposed backfit and the availability of such resources.

f. A consideration of important qualitative factors bearing on the need for the backfit at the particular facility, such as, but not limited to, operational trends, significant plant events, management effectiveness, or results of performance reports such as the Systematic Assessment of Licensee Performance.

g. A statement affirming appropriate interoffice coordination related to the proposed backfit and the plan for implementation.

h. The basis for requiring or permitting implementation on a particular schedule, including sufficient information to demonstrate that the schedules are realistic and provide adequate time for in-depth engineering, evaluation, design, procurement, installation, testing, development of operating procedures, and training of operators and other plant personnel, as appropriate. For those plants with approved integrated schedules, the integrated scheduling process can be used for implementing this step and the following two procedural steps.

Approved: August 26, 1988

i. A schedule for staff actions involved in implementation and verification of implementation of the backfit, as appropriate.

j. Importance of the proposed backfit considered in light of other safety-related activities underway at the affected facility.

k. A statement of the consideration of the proposed plant-specific backfit as a potential generic backfit.

044 Appeal Process. The appeal processes described in this section are of two types, applied to two distinctly different situations:

a. Appeal to an Office/Region to modify or withdraw a proposed backfit which has been identified, and for which a regulatory analysis has been prepared and transmitted to the licensee; or

b. Appeal to an Office/Region to reverse a denial of a prior licensee claim either that a staff position, not identified by the NRC as a backfit, is one, or that a backfit which staff believes falls within one of the exceptions from the requirement for a regulatory analysis, does not.

In the first type of situation described, licensees should address an appeal of a proposed backfit to the Office Director or Regional Administrator whose staff proposed the backfit with a copy to the EDO. The appeal should provide arguments against the rationale for imposing a backfit as presented in the staff's regulatory analysis. The Office Director or Regional Administrator shall report to the EDO within 3 weeks after receipt of the appeal concerning the plan for resolving the issue. The licensee should also be promptly and periodically informed in writing regarding the staff plans. The decision of the Office Director on an appeal of plant-specific backfit may be appealed to the EDO unless resolution is achieved at a lower management level. The EDO shall promptly resolve the appeal and shall state his reasons therefor. Summaries of all appeal meetings shall be prepared promptly, provided to the licensee, and placed in appropriate Public Document Rooms. During the appeal process, primary consideration shall be given to how and why the proposed backfit provides a substantial increase in overall protection and whether the associated costs of implementation are justified in view of the increased protection. This consideration should be made in the context of the regulatory analysis as well as any other information that is relevant and material to the proposed backfit.

In the second type of appeal situation the appeal should be addressed to, and will be decided by, the Director of the program office having responsibility for the program area relevant to the staff position, unless resolution is achieved at a lower management level. A copy of the appeal should also be sent to the Executive Director for Operations. The appeal should take into account the staff's evaluation, the licensee's response, and any other information that is relevant and material to the backfit determination. The EDO may review and may modify a decision either at his or her own initiative or at the request of the licensee. If the licensee appeals to the EDO, the EDO

Approved: August 26, 1988

shall promptly resolve the appeal and shall state the reasons therefor. Backfit claims and resultant staff determinations that are reevaluated in response
to an appeal, and that are again determined by the NRC not to be backfits, or
are excepted from the requirement for a regulatory analysis, are not to be
treated further in the context of this chapter. Such matters are to be dealt
with within the normal licensing or inspection appeal process and are not
subject to the requirements of this chapter.

045   Implementation of Backfits. Following approval of any required regulatory analysis by the appropriate Office Director or Regional Administrator,
review if any by the EDO, and issuance of the backfit to the licensee, the
licensee will either implement the backfit or appeal it. After an appeal and
subsequent final decision by the appropriate Office Director or EDO, the licensee may elect to implement a backfit resulting from the decision. If the
licensee does not elect to implement the backfit, it may be imposed by Order
of the appropriate Office Director.[3]

Implementation of plant-specific backfits will normally be accomplished on a
schedule negotiated between the licensee and the NRC. Scheduling criteria
should include the importance of the backfit relative to other safety related
activities underway, or the plant construction or maintenance planned for the
facility, in order to maintain high quality construction and operations. For
plants that have integrated schedules, the integrated scheduling process can
be used for this purpose.

A staff-proposed backfit may be imposed by Order[3] prior to completing any of
the procedures set forth in this chapter provided the NRC official authorizing
the Order determines that immediate imposition is necessary to provide adequate protection to the public health and safety or the common defense and
security. In such cases, the EDO shall be notified promptly of the action and
a documented evaluation as described in Section 042 performed, if possible, in
time to be issued with the order.

If "immediate imposition" is not necessary, staff proposed backfits shall not
be imposed, and plant construction, licensing action, or operation shall not
be interrupted or delayed by NRC actions, during the staff's evaluation and
backfit transmittal process, or a subsequent appeal process, until final action is completed under this chapter.

046   Recordkeeping and Reporting. The proposing Headquarters Office or
Regional Office shall administratively manage each proposed plant-specific
backfit using one agency recordkeeping system that provides for prompt re-
trieval of current status, planned and accomplished schedules, and ultimate
disposition. The system shall provide reference to all documents issued or
received by NRC staff relative to a plant-specific backfit, including requests, positions, statements, and summary reports. Access to make changes
to the system will be limited to those designated within each Office and
Region. Specific data required will include, but are not limited to:

---

[3]Once an Order is issued, whether or not it is immediately effective, this
chapter no longer applies and appeals are governed by the procedures in 10
CFR Part 2, Subpart B.

Approved: August 26, 1988

a.   Licensee and facility affected.

b.   Whether a backfit is identified by staff or by a licensee.

c.   Identification and description of the document that either transmits a staff-identified backfit or a licensee request for consideration of a licensee-identified backfit.

d.   Substance of the backfit issue.

e.   In the case of a licensee-identified backfit, the dates (predicted and completed) that determinations are made as to whether or not a staff position qualifies as a backfit, the substance of the determination, and the organization and official responsible for making the determination.

f.   A brief description of what action is pending, and the officials responsible to complete the action.

g.   Action closing date, to include a description of licensee or staff action and date of agreement or order to implement; responsible officials and organization for each action.

047   Exceptions. Nothing in this chapter shall be interpreted as authorizing or requiring the staff to make plant-specific backfits or assessments for generic backfits that are, or have been, subject to review by the CRGR and approval by the EDO, or for generic backfits approved prior to November 1981, unless the EDO determines that significant plant-specific issues were not considered during the prior reviews or the EDO authorizes a deviation under Section 031.

048   References.

a.   NUREG/BR-0058, Rev. 1, May 1984, "Regulatory Analysis Guidelines of the U.S. Nuclear Regulatory Commission"

b.   NUREG/CR-3568, December 1983, "A Handbook for Value-Impact Assessment"

c.   NUREG/CR-3971, October 1984, "A Handbook for Cost Estimating"

d.   Revision of Backfit Rule, Code of Federal Regulations, 53 FR 20603 (June 6, 1988)

0514-05   DEFINITIONS

051   Licensee. Except where defined otherwise, the word licensee as used in this chapter shall mean that person that holds a license to operate a nuclear power plant, or a construction permit to build a nuclear power plant, or a Preliminary Design Approval, Final Design Approval, or Design Certification for a Standardized Plant Design.

Approved: August 26, 1988

052 Plant-Specific Backfit. Backfitting is defined as the modification of or addition to systems, structures, components, or design of a facility; or the design approval or manufacturing license for a facility; or the procedures or organization required to design, construct or operate a facility; any of which may result from a new or amended provision in the Commission rules or the imposition of a regulatory staff position interpreting the Commission rules that is either new or different from a previously applicable staff position after certain specified dates. Backfitting is "plant-specific" when it involves the imposition of a position that is unique to a particular plant.

It should be noted that to be a plant-specific backfit a staff position must meet conditions involving both (a) the substance of the elements of a proposed staff position and (b) the time of the identification of the staff position:

a.   A staff position may be a proposed backfit if it would cause a licensee to change the design, construction or operation of a facility from that consistent with already applicable regulatory staff positions. Applicable regulatory staff positions are described in Section 053.

b.   A staff position as described in (a) above is a proposed backfit if it is first identified to the licensee after certain important design, construction or operation milestones, involving NRC approvals of varying kind, has been achieved. Those times after which a new or revised staff position will be considered a backfit are as follows:

(1)   After the date of issuance of the construction permit for the facility (for facilities having construction permits issued after May 1, 1985);

(2)   After 6 months before the date of docketing of the OL application for the facility (for facilities having construction permits issued before May 1, 1985);

(3)   After the date of issuance of the operating license for the facility (for facilities having an operating license on May 1, 1985);

(4)   After the date of issuance of the design approval under 10 CFR 50, Appendix M, N or 0.

NOTE: The EDO directives embodied in chapter NRC-0514 are effective as of July 6, 1988.

053 Applicable Regulatory Staff Positions. Applicable regulatory staff positions are those already specifically imposed upon or committed to by a licensee at the time of the identification of a plant-specific backfit, and are of several different types and sources:

a.   Legal requirements such as in explicit regulations, orders, plant licenses (amendments, conditions, technical specifications). Note

Approved: August 26, 1988

that some regulations have update features built in, as for example, 10 CFR 50.55a, Codes and Standards. Such update requirements are applicable as described in the regulation.

b. Written commitments such as contained in the FSAR, LERs, and docketed correspondence, including responses to Bulletins, responses to Generic Letters, Confirmatory Action Letters, responses to Inspection Reports, or responses to Notices of Violation.

c. NRC staff positions[4] that are documented, approved, explicit interpretations of the more general regulations, and are contained in documents such as the SRP, Branch Technical Positions, Regulatory Guides, Generic Letters, and Bulletins; and to which a licensee or an applicant has previously committed to or relied upon. Positions contained in these documents are not considered applicable staff positions to the extent that staff has, in a previous licensing or inspection action, tacitly or explicitly excepted the licensee from part or all of the position.[5]

---

[4]Requirements may be imposed by rule or order. Staff interpretations such as examples of acceptable ways to meet requirements are not requirements in and of themselves.

[5]Imposition of a staff position from which a licensee has previously been excepted is a backfit.

Approved: August 26, 1988

CONTENTS

Guidance for Making Backfit Determinations   Page

A.   General... 1
B.   Licensing.. 1

     1.   Standard Review Plan.. 1
     2.   Regulatory Guides... 2
     3.   Plant-Specific Orders... 3

C.   Inspection and Enforcement... 3

     1.   Inspections... 3
     2.   Notice of Violations.. 4
     3.   Bulletins... 4
     4.   Reanalysis of Issues.. 4

i   Approved: August 26, 1988

Guidance for Making Backfit Determinations

A. General

In this section selected regulatory activities and documents are discussed in order to enable members of the NRC staff and the regulated industry to better understand the conditions under which a staff position may be viewed as a plant-specific backfit. It is important to understand that the necessity for making backfit determinations should not inhibit the normal informal dialogue between the technical reviewer or inspector and the licensee. The intent of this process is to manage backfit imposition, not to quell it. The discussion in this appendix is intended to aid in identifying backfits in accordance with the principles and the practices that should be implemented by all staff members. This appendix is not intended to be an exhaustive, comprehensive workbook in which can be found a parallel example for each situation that may arise. As is evident from the definitions in Section 05 of this chapter, a plant-specific backfit has the elements of a change from an already applicable staff position where an applicable staff position is defined as that established before certain defined milestones in the affected plant's licensing history. There will be some judgment necessary to determine whether a staff position would cause a licensee to change the design, construction or operation of a facility. In making this determination, the fundamental question is whether the staff's action is directing, telling, or coercing, or is merely suggesting or asking the licensee to consider a staff proposed action.

Actions proposed by the licensee are not backfits under this chapter even though such actions may result from normal discussions between staff and licensee concerning an issue, and even though the change or additions may meet the definitions of Section 0514-052 and 0514-053.

B. Licensing

1. Standard Review Plan (SRP) - The SRP delineates the scope and depth of staff review of licensee submittals associated with various licensing activities. It is a definitive NRC staff interpretation of measures which, if taken, will satisfy the requirements of the more generally stated, legally binding body of regulations, primarily found in Title 10 CFR. Since October 1981, changes to the SRP are to have been reviewed and approved through a generic review process involving the Committee to Review Generic Requirements (CRGR), and the extent to which the changes apply to classes of plants is defined. Consequently, application of a current SRP in a specific operating license (OL) review generally is not a plant-specific backfit, provided the SRP was effective 6 months prior to the start of the OL review. Asking an applicant for an operating license questions to clarify staff understanding of proposed actions, in order to determine whether the actions will meet the intent of the SRP, is not considered a backfit.

On the other hand, using acceptance criteria more stringent than those contained in the SRP or taking positions more stringent than

1   Approved: August 26, 1988

in addition to those specified in the SRP, whether in writing or orally, is a plant-specific backfit. During meetings with the licensee, staff discussion or comments regarding issues and licensee actions volunteered which are in excess of the criteria in the SRP generally do not constitute plant-specific backfits; however, if the staff implies or suggests that a specific action in excess of already applicable staff positions is the only way for the staff to be satisfied, the action is considered a plant-specific backfit whether or not the licensee agrees to take such action. However, the staff should recognize that a verbally implied or suggested action should not be accepted by a licensee as an NRC position of any kind, backfit or not; only written and authoritatively approved position statements should be taken as NRC positions.

Application of an SRP to an operating plant after the license is granted generally is considered a backfit unless the SRP was approved specifically for operating plant implementation and is applicable to such operating plant. It is important to note, however, that in order to issue an amendment to a license, there must be a current finding of compliance with regulations applicable to the amendment. As a specific example, review of a plant owner's application for a license amendment to authorize installation and operation of a new reactor core, commonly called a "reload application," may necessitate review of new fuel designs or new thermal-hydraulic correlations and associated operating limits. Such changes that are clearly advances in design or operation may involve new or unreviewed safety issues, and may warrant review to SRP criteria which were approved subsequent to initial license issuance to the licensee. This is not considered a backfit. However, such review to newer SRP revisions is not necessarily required to determine current compliance with regulations. Licensee-proposed revisions in design or operation that raise staff questions only about potential reduced margins of safety as defined in the basis for any technical specification should be reviewed by reanalysis of the same accident sequences and associated assumptions as analyzed in the FSAR for the initial license issuance.

During reload reviews, staff-proposed positions with regard to technical matters not related to the changes proposed by a licensee shall be considered backfits.

2. Regulatory Guides - As part of the generic review process pursuant to the CRGR Charter, it is decided which plants or groups of plants should be affected by new or modified Regulatory Guide provisions. Such implementation is therefore not governed by the plant-specific backfit procedures. However, any staff proposed plant-specific implementation of a Regulatory Guide provision, whether orally or in writing, for a plant not encompassed by the generic implementation determination is considered a plant-specific backfit. A staff action with respect to a specific licensee that expands on, adds to, or modifies a generically approved regulatory guide, such that the position taken is more demanding than intended in the generic positions, is a plant-specific backfit.

Approved: August 26, 1988   2

3. Plant-Specific Orders - An order issued to cause a licensee to take actions which are not otherwise applicable regulatory staff positions is a plant-specific backfit. As described in Section 0514-045 of this chapter, an order effecting immediate imposition of a backfit may be issued prior to completing any of the procedures set forth in this chapter provided that the appropriate Office Director determines that immediate imposition is necessary.

    An order issued to confirm a licensee commitment to take specific action even if that action is in excess of previously applicable staff positions, is not a plant-specific backfit provided the commitment was not obtained by the staff with the expressed or implied direction that such a commitment was necessary to gain acceptance in the staff review process. Discussion or comments by the NRC staff identifying deficiencies observed, whether in meetings or written reports, do not constitute backfits. Definitive statements to the licensee directing a specific action to satisfy staff positions are backfits unless the action is an explicit and already applicable regulatory staff position.

C. Inspection and Enforcement

1. Inspections - NRC inspection procedures govern the scope and depth of staff inspections associated with licensee activities such as design, construction, and operation. As such, they define those items the staff is to consider in its determination of whether the licensee is conducting its activities in a safe manner. The conduct of inspections establishes no new staff positions for the licensee and is not a plant-specific backfit.

    Staff statements to the licensee that the contents of an NRC inspection procedure are positions that must be met by the licensee constitute a plant-specific backfit unless the item is an applicable regulatory staff position. Discussion or comment by the NRC staff regarding deficiencies observed in the licensee conduct of activities, whether in meetings or in written inspection reports, do not constitute backfits, unless the staff suggests that specific corrective actions different from previous applicable regulatory staff positions are the only way to satisfy the staff. In the normal course of inspecting to determine whether the licensee's activities are being conducted safely, inspectors may examine and make findings in specific technical areas wherein prior NRC positions and licensee commitments do not exist. Examination of such areas and making findings is not considered a backfit. Likewise, discussion of findings with the licensee is not considered a backfit. If during such discussions, the licensee agrees that it is appropriate to take action in response to the inspector's findings, such action is not a backfit provided the inspector does not indicate that the specific actions are the only way to satisfy the staff. On the other hand, if the inspector indicates that a specific action must be taken, such action is a backfit unless it constitutes an applicable regulatory staff position. Further, if the licensee decides

3   Approved: August 26, 1988

to claim that the inspector's findings are a backfit, then the staff must decide whether they are a backfit under this chapter.

For example, if the licensee commits to ANSI-N18.7 in the SAR and the inspector finds the licensee's implementing procedures do not contain all the elements required by ANSI-N18.7, telling the licensee he must take action to include all the elements in the implementing procedures is not a backfit. If the inspector finds the licensee has included all the required elements of ANSI-N18.7, but has not included certain of the optional elements in the implementing procedures, inspector discussion with the licensee regarding the merits of including the optional elements is not a backfit. On the other hand, if the inspector tells the licensee that the implementing procedures must include any or all of the optional elements in order to satisfy the staff, inclusion of such elements is a backfit, whether or not agreed to by the licensee.

2. Notice of Violations (NOV) - a NOV requesting description of a licensee's proposed corrective action is not a backfit. The licensee's commitments in the description of corrective action are not backfits. A request by the staff for the licensee to consider some specific action in response to an NOV is not a backfit. However, if the staff is not satisfied with the licensee's proposed corrective actions and requests that the licensee take additional actions, those additional actions (whether requested orally or in writing) are a backfit unless they are an applicable regulatory staff position.

    Discussions during enforcement conferences and responses to the licensees requests for advise regarding corrective actions are not backfits; however, definitive statements to the licensee directing a specific action to satisfy staff positions are backfits, unless the action is an explicit applicable regulatory staff position.

3. Bulletins - Bulletins and resultant actions requested of licensees undergo the general review process pursuant to the CRGR Charter. Therefore, in general, it is not necessary to apply the plant-specific backfit process to the actions requested in a Bulletin. However, if the staff expands the action requested by a Bulletin during its application to a specific licensee, such expansion is considered a plant-specific backfit.

4. Reanalysis of Issues - Throughout plant lifetime, many individuals on the NRC staff have an opportunity to review the requirements and commitments incumbent upon a licensee. Undoubtedly, there will be occasions when a reviewer concludes the licensee's program in a specific area does not satisfy a regulation, license condition or commitment. In the case where the staff previously accepted the licensee's program as adequate, any staff specified change in the program would be classified as a backfit.

Approved: August 26, 1988   4

For example, in the case of an NTOL, once the SER is issued signi-fying staff acceptance of the programs described in the SAR, the licensee should be able to conclude that his commitments in the SAR satisfy the NRC requirements for a particular area. If the staff was to subsequently require that the licensee commit to additional action other than that specified in the SAR for the particular area, such action would constitute a backfit.

A somewhat different situation exists when the licensee has made a submittal committing to a specific course of action to meet an applica-ble position, and the staff has not yet responded, and therefore has not indicated that the commitment is or is not sufficient to meet the applicable position. Subsequent staff action, which must be taken within a reasonable time not delaying the applicant's implemen-tation plans, to cause the licensee to meet the applicable regulatory staff position is not a backfit. If the licensee has moved ahead in the intervening time to implement that which the licensee proposed to do in its submittal and the staff has failed to provide a timely response, then the staff position may be considered a backfit. Thus, if a licensee has implemented a technical resolution intended to meet an applicable regulatory staff position, and staff for an extended period simply allows the licensee resolution to stand with tacit accep-tance indicated by non-action on the part of NRC, then a subsequent action to change the licensee's design, construction, or operation is a backfit.

5   Approved: August 26, 1988

APPENDIX E

BACKGROUND INFORMATION FOR CRGR REVIEW
OF GI–70 AND GI–94 RESOLUTIONS

Background Information for CRGR Review
of GI-70 and GI-94 Resolutions

The following information is provided in the format specified in Section IV B(i) through IV B(ix) of Revision 4 of the CRGR Charter, dated April 1987. For each item, the request for information is given followed by a discussion of the response or a reference to where the information is provided.

(i)   The proposed generic requirement or staff position as it is proposed to be sent out to licensees.

The proposed generic resolution is set forth in the proposed 10CFR50.54(f) generic letter <u>89-XX</u> (see Enclosure 2).

(ii)   Draft staff papers or other underlying staff documents supporting the requirements or staff positions. (A copy of all materials referenced in the document shall be made available upon request to the CRGR staff. Any committee member may request CRGR staff to obtain a copy of any referenced material for his or her use.)

The relevant technical information for GI-70 is contained in NUREG-1316 (Enclosure 3) and related contractor reports and other references listed therein. The relevant technical information for GI-94 is contained in NUREG-1326 (Enclosure 10) and related contractor reports and other references listed therein. Copies of any references will be provided upon request.

(iii)   Each proposed requirement or staff position shall contain the sponsoring office's position as to whether the proposal would increase requirements or staff positions, implement existing requirements or staff positions, or would relax or reduce existing requirements or staff positions.

Technical findings related to the resolution of GI-70 are contained in NUREG-1316 (Enclosure 3). Technical findings related to the resolution of GI-94 are contained in NUREG-1326 (Enclosure 10). These findings have been incorporated in the proposed generic letter (Enclosure 2). They represent the final staff position on GI-70 and GI-94 and for certain operating PWR plants are additional requirements. For certain recently licensed operating plants and certain plants currently under active construction the GI-70 technical findings do not represent additional requirements. The GI-94 technical findings do not represent additional requirements for Babcock and Wilcox plants.

(iv)   The proposed method of implementation along with the concurrence (and any comments) of OGC on the method proposed.

OGC has no legal objection to the proposed action in Generic Letter 89-XX (Enclosure 2). All OGC comments have been incorporated in the proposed generic letter.

(v)   Regulatory analyses generally conforming to the directives and guidance of NUREG/CR-0058 and NUREG/CR-3568.

Regulatory analyses related to the resolution of GI-70 are contained in Section 5 of NUREG-1316 (Enclosure 3). The Regulatory Analysis for the resolution of GI-94 is provided in NUREG-1326 (Enclosure 10).

(vi)   Identification of the category of reactor plants to which the generic requirement or staff position is to apply (that is, whether it is to apply to new plants only, new OLs only, OLs after a certain date, OLs before a certain date, all OLs, all plants under construction, all plants, all water reactors, all PWRs only, some vendor types, some vintage types such as BWR 6 and 4, jet pump and nonjet pump plants, etc.)

The proposed Generic Letter 89-XX (Enclosure 2) will be sent to PWRs and is applicable to all operating plants and future plants including those currently under construction. However, certain recently licensed operating plants and certain plants currently under active construction already satisfy the GI-70 recommendations of the generic letter, and certain plants are not impacted by the GI-70 recommendations (CE plants without PORVs). With respect to the GI-94 recommendation, Babcock and Wilcox plants are not impacted.

(vii) For each such category of reactor plants, an evaluation which demonstrates how the action should be prioritized and scheduled in light of other ongoing regulatory activities. The evaluation shall document for consideration information available concerning any of the following factors as may be appropriate and any other information relevant and material to the proposed action:

Potential improvements to PORVs and block valves should be prioritized and scheduled in conjunction with ongoing regulatory activities such as; review of inservice testing programs of valves in conformance with Section XI of the ASME Code. Review of potential modifications to technical specifications for low-temperature overpressure protection should be prioritized in conjunction with these same activities.

(a)   Statement of the specific objectives that the proposed action is designed to achieve;

The objectives that the proposed actions are designed to achieve are to increase the reliability of PORVs and block valves to provide assurance they will function as required, and to provide additional assurance that LTOP systems will be available when required.

(b) General description of the activity that would be required by the licensee or applicant in order to complete the action;

With respect to GI-70, for operating plants when PORVs and the associated block valves are used for any of the safety functions discussed in Section 2.1 of NUREG-1316 the activity that would be required by the licensee consists of the following actions:*

(1) Include PORVs and block valves in the operational quality assurance program that is in compliance with 10 CFR 50, Appendix B.

(2) Provide a maintenance/refurbishment program for PORVs and block valves.

(3) Testing in accordance with Section XI of the ASME Code for PORVs and block valves. Additional testing for PORV block valves will be included in the expanded MOV test program discussed in NRC Generic Letter 89-XX, "Safety-Related Motor Operated Valve Testing and Surveillance" dated (Later).

(4) Modify the limiting conditions of operation of PORVs and block valves in the technical specifications for Modes 1, 2, and 3 as contained in Attachments A-1, A-2, and A-3 of Enclosure A to the proposed Generic Letter 89-XX (Enclosure 2).

For future PWR plants and those currently under construction when PORVs and the associated block valves are used for any of the safety functions discussed in Section 2.1 of NUREG-1316, these components should be classified as safety related and a minimum of two PORVs and two block valves installed. Plants currently under active construction meet these recommendations.

With respect to GI-94, Combustion Engineering and Westinghouse PWRs should modify the current plant Technical Specifications for the Overpressure Protection System to assure both channels are operable in Modes 5 and 6, especially when water-solid as contained in Enclosure B and its attachments to proposed Generic Letter 89-XX (Enclosure 2). Revisions to the plant cooldown and heatup (or filling and venting) procedures are also recommended. In addition, verification that administrative controls and procedures regarding the LTOP design basis analyses have been implemented is also recommended.

(c) Potential change in the risk to the public from the accidental offsite release of radioactive material;

---

*Certain recently licensed operating plants already satisfy these requirements.

GI-70   Contractor analysis showed only a small change in risk based on Indian Point 3 and Oconee PRA's. However, NUREG/CR-5230 showed that feed and bleed provides a significant reduction in core melt probability for four representative plants. The proposed actions would enhance feed and bleed capability. Even if only a fraction of the core melt reduction indicated in NUREG/CR-5230 is achieved, this would result in a substantial reduction in risk to the public.

GI-94   The estimated total dose reduction is 14,500 person-rem over the remaining license life of the PWRs impacted by the proposed resolutions.

(d)   Potential impact on radiological exposure of facility employees and other onsite workers.

For GI-70, it is estimated that there would be little or no increase in exposure because:

(1)   Most surveillance testing would be performed remotely in situ.

(2)   Exposure resulting from orderly planned maintenance activities is considered unlikely to result in exposure levels any higher than those resulting from unplanned major repairs after valves malfunction in service.

No additional exposure to facility employees or other onsite workers is expected for the proposed resolution of GI-94.

(e)   Installation and continuing costs associated with the action, including the cost of facility downtime or the cost of construction delay;

The present worth of the utility cost impact for GI-70 for operating PWR plants with two PORVs and two block valves is $127,200 for items (b)1 through (b)4 discussed above. However, this cost will be more than offset by the savings from less outage time because of PORV and block valve problems. This work would be accomplished during scheduled refueling/maintenance outages as a part of existing plant programs.

The net present value of the estimated replacement power cost resulting from the proposed resolution of GI-94 is estimated to be $2,000 per plant, assuming a 5% discount rate and a 24 year average remaining lifetime for the plants impacted by GI-94. The average annual utility cost is estimated to be $145.00.

(f)   The potential safety impact of changes in plant or operational complexity, including the relationship to proposed and existing regulatory requirements and staff positions;

For a certain number of operating plants the proposed revision to the technical specifications may be more restrictive. The proposed recommendations for GI-70 are expected to increase the reliability of PORVs in Modes 1, 2 and 3 and therefore overall plant safety.

The proposed recommendations for GI-94 are expected to increase the availability of the LTOP systems in Modes 5 and 6 (especially when water-solid) and therefore overall plant safety.

(g) The estimated resource burden on the NRC associated with the proposed action and the availability of such resources;

The estimated resource burden on the NRC is minimal, costs are estimated in Section 5.5 of the Regulatory Analysis in NUREG-1316 (Enclosure 3) for GI-70, and in Section 5.1.2 of the Regulatory Analysis in NUREG-1326 (Enclosure 10) for GI-94.

(h) The potential impact of differences in facility type, design or age on the relevancy and practicality of the proposed action;

The potential impact of the proposed actions for GI-70 on PWRs with PORVs will vary from none to moderate. That is, on plants that received an OL since 1984 there would be no impact as these plants in general have safety grade PORVs and block valves. On older PWRs with PORVs the impact will be variable depending on the degree of compliance with items (b)1 through (b)4 discussed above. CE plants without PORVs are not impacted by the proposed resolution for GI-70

The potential impact of the proposed actions for GI-94 are not expected to be different based on facility type, design or age. However, Babcock and Wilcox PWRs are not impacted by the proposed resolution for GI-94.

(i) Whether the proposed action is interim or final, and if interim, the justification for imposing the proposed action on an interim basis.

The proposed actions are final with respect to the resolution of GI-70 and GI-94.

APPENDIX F

SAMPLE BACKFIT DISCUSSION
(TAKEN FROM NRC BULLETIN 90-01)

Backfit Discussion

The objective of the actions requested in this bulletin is to ensure that transmitter failures due to loss of fill-oil are promptly detected. Loss of fill-oil may result in a transmitter not performing its intended safety function.

The actions requested in this bulletin represent new staff positions and thus, this request is considered a backfit in accordance with NRC procedures. Because established regulatory requirements exist but were not satisfied, this backfit is to bring facilities into compliance with existing requirements. Therefore, a full backfit analysis was not performed. An evaluation of the type discussed in 10 CFR.109(a)(6) was performed, including a statement of the objectives of and reasons for the actions requested and the basis for invoking the compliance exception. It will be made available in the Public Document Room with the minutes of the 179th meeting of the Committee to Review Generic Requirements.

NRC FORM 335
(2-89)
NRCM 1102,
3201, 3202

U.S. NUCLEAR REGULATORY COMMISSION

BIBLIOGRAPHIC DATA SHEET

(See instructions on the reverse)

1. REPORT NUMBER
(Assigned by NRC. Add Vol., Supp., Rev., and Addendum Numbers, if any.)

NUREG-1409

2. TITLE AND SUBTITLE

Backfitting Guidelines

3. DATE REPORT PUBLISHED

MONTH	YEAR
July	1990

4. FIN OR GRANT NUMBER

5. AUTHOR(S)

Allison, Dennis P.
Conran, James H.
Trottier, Cheryl A.

6. TYPE OF REPORT

7. PERIOD COVERED *(Inclusive Dates)*

8. PERFORMING ORGANIZATION — NAME AND ADDRESS *(If NRC, provide Division, Office or Region, U.S. Nuclear Regulatory Commission, and mailing address; if contractor, provide name and mailing address.)*

Office for Analysis and Evaluation of Operational Data
U.S. Nuclear Regulatory Commission
Washington, DC 20555

9. SPONSORING ORGANIZATION — NAME AND ADDRESS *(If NRC, type "Same as above"; if contractor, provide NRC Division, Office or Region, U.S. Nuclear Regulatory Commission, and mailing address.)*

Same as above.

10. SUPPLEMENTARY NOTES

11. ABSTRACT *(200 words or less)*

The backfitting process is the process by which the U.S. Nuclear Regulatory Commission (NRC) decides whether to issue new or revised requirements or staff positions to licensees of nuclear power reactor facilities. Requirements for proper justification of backfits and information requests are provided by two NRC rules (Title 10, Code of Federal Regulations, Sections 50.109 and 50.54(f)). NRC procedures include the charter of the committee to Review Generic Requirements, NRC Manual Chapter 0514, and individual office procedures. Three types of backfits are recognized. Cost-justified substantial safety improvements require backfit analyses and findings of (1) substantial safety improvement and (2) justified costs. Compliance exceptions and adequate protection exceptions do not require findings of substantial safety improvements and costs are not considered. However, they are still backfits and require documented evaluations to support the use of the exceptions. Information requests (as opposed to backfits) require an analysis of the burden to be imposed to ensure that they are justified in view of the potential safety significance of the information requested.

12. KEY WORDS/DESCRIPTORS *(List words or phrases that will assist researchers in locating the report.)*

backfit
backfit analysis
information request
regulatory analysis
cost-benefit analysis
generic communication
new requirement

13. AVAILABILITY STATEMENT

Unlimited

14. SECURITY CLASSIFICATION

(This Page)

Unclassified

(This Report)

Unclassified

15. NUMBER OF PAGES

16. PRICE

www.ingramcontent.com/pod-product-compliance
Lightning Source LLC
Chambersburg PA
CBHW080306180526
45167CB00006B/2694